RUSSIA'S ARCTIC

RUSSIAN SHORTS

Russian Shorts is a series of thought-provoking books published in a slim format. The Shorts books examine key concepts, personalities, and moments in Russian historical and cultural studies, encompassing its vast diversity from the origins of the Kievan state to Putin's Russia. Each book is intended for a broad range of readers, covers a side of Russian history and culture that has not been well-understood, and is meant to stimulate conversation.

Series Editors:
Eugene M. Avrutin, Professor of Modern European Jewish History, University of Illinois, USA
Stephen M. Norris, Professor of History, Miami University, USA

Editorial Board:
Edyta Bojanowska, Professor of Slavic Languages and Literatures, Yale University, USA
Ekaterina Boltunova, Associate Professor of History, Higher School of Economics, Russia
Eliot Borenstein, Professor of Russian and Slavic, New York University, USA
Melissa Caldwell, Professor of Anthropology, University of California, Santa Cruz, USA
Choi Chatterjee, Professor of History, California State University, Los Angeles, USA
Robert Crews, Professor of History, Stanford University, USA
Dan Healey, Professor of Modern Russian History, University of Oxford, UK
Polly Jones, Professor of Russian, University of Oxford, UK
Paul R. Josephson, Professor of History, Colby College, USA
Marlene Laruelle, Research Professor of International Affairs, George Washington University, USA
Marina Mogilner, Associate Professor, University of Illinois at Chicago, USA

Willard Sunderland, Henry R. Winkler Professor of Modern History, University of Cincinnati, USA

Published Titles
Pussy Riot: Speaking Punk to Power, Eliot Borenstein
Memory Politics and the Russian Civil War: Reds Versus Whites, Marlene Laruelle and Margarita Karnysheva
Russian Utopia: A Century of Revolutionary Possibilities, Mark D. Steinberg
Racism in Modern Russia: From the Romanovs to Putin, Eugene M. Avrutin
Meanwhile, in Russia: Russian Internet Memes and Viral Video, Eliot Borenstein
Ayn Rand and the Russian Intelligentsia: The Origins of an Icon of the American Right, Derek Offord
The Multiethnic Soviet Union and Its Demise, Brigid O'Keeffe
Nuclear Russia: The Atom in Russian Politics and Culture, Paul Josephson
The Afterlife of the "Soviet Man": Rethinking Homo Sovieticus, Gulnaz Sharafutdinova
The History of Birobidzhan: Building a Soviet Jewish Homeland in Siberia, Gennady Estraikh
The Soviet Gulag: History and Memory, Jeffrey S. Hardy
Why We (Still) Need Russian Literature: Tolstoy, Dostoevsky, Chekhov and Others, Angela Brintlinger
Russian Food since 1800: Empire at Table, Catriona Kelly
Jews under Tsars and Communists: The Four Questions, Robert Weinberg
Gulag Fiction: Labour Camp Literature from Stalin to Putin, Polly Jones
Russian Culture under Putin, Eliot Borenstein
How Russia Got Big: A Territorial History, Paul W. Werth
Soviet Internment: Memory, Nostalgia and the POW Experience, Maria Cristina Galmarini
Racism in Modern Russia – Revised Edition: From the Romanovs to Putin, Eugene M. Avrutin

Russia's Arctic: Climate Change, Domestic Policy, and Geopolitics, Marlene Laruelle and Jean Radvanyi

Upcoming Titles
Russia's History Painters: Vasily Surikov, Viktor Vasnetsov, and the Remaking of the Past, Stephen M. Norris
A Social History of the Russians and Their Army since 1690, Roger R. Reese
Black Encounters with the Soviet Union: Hope Meets Promise, Maxim Matusevich
The Invention of Russian Time, Andreas Schönle
The Tolstoy Marriage: A Literary History, Ani Kokobobo

RUSSIA'S ARCTIC

CLIMATE CHANGE, DOMESTIC POLICY, AND GEOPOLITICS

Marlene Laruelle and Jean Radvanyi

BLOOMSBURY ACADEMIC
LONDON • NEW YORK • OXFORD • NEW DELHI • SYDNEY

BLOOMSBURY ACADEMIC
Bloomsbury Publishing Plc, 50 Bedford Square, London, WC1B 3DP, UK
Bloomsbury Publishing Inc, 1359 Broadway, New York, NY 10018, USA
Bloomsbury Publishing Ireland, 29 Earlsfort Terrace, Dublin 2,
D02 AY28, Ireland

BLOOMSBURY, BLOOMSBURY ACADEMIC and the Diana logo are
trademarks of Bloomsbury Publishing Plc

First published in Great Britain 2022
This edition published 2026

Copyright © Marlene Laruelle and Jean Radvanyi, 2026

Marlene Laruelle and Jean Radvanyi have asserted their right under the
Copyright, Designs and Patents Act, 1988, to be identified as Authors of
this work.

Series design by Tjaša Krivec
Cover image: A view of the "Arctic Trefoil," or three-lobed leaf,
military base on the island of Alexandra Land,
which is part of the Franz Josef Land archipelago, on May 17, 2021.
© Photo by Maxime POPOV / AFP, Getty Images

All rights reserved. No part of this publication may be: i) reproduced or
transmitted in any form, electronic or mechanical, including photocopying,
recording or by means of any information storage or retrieval system without
prior permission in writing from the publishers; or ii) used or reproduced in
any way for the training, development or operation of artificial intelligence
(AI) technologies, including generative AI technologies. The rights holders
expressly reserve this publication from the text and data mining exception as
per Article 4(3) of the Digital Single Market Directive (EU) 2019/790.

Bloomsbury Publishing Plc does not have any control over, or responsibility
for, any third-party websites referred to or in this book. All internet addresses
given in this book were correct at the time of going to press. The author and
publisher regret any inconvenience caused if addresses have changed or sites
have ceased to exist, but can accept no responsibility for any such changes.

A catalogue record for this book is available from the British Library.

A catalog record for this book is available from the Library of Congress.

ISBN: HB: 978-1-3505-5257-9
PB: 978-1-3505-5256-2
ePDF: 978-1-3505-5258-6
eBook: 978-1-3505-5259-3

Typeset by Newgen KnowledgeWorks Pvt. Ltd., Chennai, India
Printed and bound in Great Britain

For product safety related questions contact productsafety@bloomsbury.com.

To find out more about our authors and books visit www.bloomsbury.com
and sign up for our newsletters.

CONTENTS

List of Illustrations — viii

Introduction — 1

1 A Strategic Facade Subject to Increased Attention — 5

2 A Major Diplomatic and Security Space for Moscow — 33

3 Successes and Challenges of Arctic Economic Transformations — 55

4 What Development for the Far North? — 75

Conclusion — 97

Arctic Chronology — 101
Notes — 107
Selected Bibliography — 135
Index — 145

ILLUSTRATIONS

Figures

3.1 The LNG port of Sabetta on the Yamal Peninsula   63

4.1 Norilsk city epitomizing Arctic urban infrastructure
© Marlene Laruelle, 2015   79

Graph

1.1 Total traffic of the Northern Sea Route (millions of tons, coastal and transit)   30

Maps

1.1 Borders and territorial claims in the Russian Arctic   16

2.1 Economic and strategic development in the Russian Arctic   48

Tables

4.1 Major Cities of the Russian Arctic by Population, 1990–2024   81

4.2 Population of Russian Arctic Regions on January 1 (in Thousands, 1990–2020)   85

4.3 Migration Flows by Region (Number of Individuals) in 2020   86

INTRODUCTION

Can the Arctic become one of the planet's hot spots? In light of the prolonged war initiated by Russia in Ukraine, this hypothesis is being revisited as an extension of the tensions observed in the Baltic Sea. The Arctic Ocean and its continental margins have always exerted a great fascination. All maritime powers, whether bordering the Arctic or not, have competed to reach the pole or gain control of the two mythical passages, the Northwest Passage off Canada and the Northeast Passage along the Russian coast beyond Scandinavia, making these dangerous seas a contested strategic space since the early twentieth century.

The sensitivity of this area to international tensions predates the Russia-Ukraine War. It is worth recalling that the Arctic archipelago of Novaya Zemlya hosted one of the Soviet Union's nuclear test sites between 1955 and 1990, and the Russian Northern Fleet is headquartered in Severomorsk, north of Murmansk. Both Russians and Westerners have conducted numerous naval maneuvers in the area, and one may remember the *Kursk* nuclear submarine disaster in August 2000, which tarnished President Vladimir Putin's first months in power and motivated him to refinance Russia's decaying armed and navy forces.

For today's Russia, the Arctic is now returning to the Cold War atmosphere of hostilities with the North Atlantic Treaty Organization (NATO). The Russian Arctic borders two NATO member states: Norway to the west and the United States across the Bering Strait to the east. The accession of Sweden and Finland to the alliance, a process begun in May 2022, further alters the fragile balance established in 1996 when the Arctic Council was created: all the Arctic states except Russia are now NATO members. On the Western side as well, the fact that Russia controls 53 percent of the circumpolar territories means that this neighbor cannot be ignored, no matter how conflictual the relationship may be. Donald Trump's reelection and his claims over Greenland have highlighted that Russia may not be the only country seeking power projection in the Arctic. Yet, the Trump-Putin meeting on August 15, 2025, in Alaska has also sent signals that both countries consider the Arctic as a place where, at least symbolically, a modest form of reset could take place, with some shared interests by both presidents.

The strategic importance of the region has taken on a whole new dimension in recent decades with two parallel developments. On the one hand, since the 2000s, there has been a proliferation of military bases, airstrips, radar stations, and ports refurbished to serve dual purposes on both sides, Western and Russian. The Arctic's demilitarized status envisioned by Gorbachev and European leaders after the end of the USSR, akin to that which still holds in Antarctica, has become nothing more than a pious wish in the north. On another level—though the two changes are interconnected—there has been a rapid development of the Northern Sea Route (NSR), primarily driven by the growth in the extraction of hydrocarbon deposits in the far north of Siberia and neighboring offshore areas. The exploitation of gas fields and liquefaction plants on the Yamal Peninsula has led to an increase in tanker flows both westward to Murmansk and eastward to new Asian clients.

However, a curious race, strewn with multiple obstacles, seems to be underway in the Russian Arctic. On the one hand, there are the paradoxical effects of climate change. Can we anticipate a time when navigation conditions at very high latitudes of the Arctic Ocean (traditionally completely covered by sea ice year-round) will allow avoiding Russian territorial waters with sufficient safety? The Northeast Passage would then become an international route without Russia being able to demand control or fees. At the same time, thawing permafrost is posing increasingly alarming technical and economic challenges to the maintenance of Russian towns and old and new industrial facilities. Moreover, will it be possible to mitigate the disastrous effects of storms, tornadoes, floods, or fires in the taiga, which tend to multiply in the sub-Arctic zone?

On an entirely different level, another race is unfolding between Western states, which are imposing increasingly broad sanctions on Russia specifically targeting Arctic hydrocarbon development projects, and a Russian government striving to circumvent or mitigate their effects. This race plays out on all fronts: delivery of technical equipment that the Russians cannot supply themselves, delivery of spare parts, boycotts of oil and gas purchases, and banking circuits linked to extraction projects or the commercial viability of production. After each new round of sanctions, Russia seeks countermeasures by relying on its allies, most of whom are Asian.

For Moscow, all these Arctic issues are vital, both economically and in terms of foreign policy and national sovereignty. The Arctic has rightly been described in official Russian documents since the 2000s as the energy heart of the country. It was first mentioned in the government's Foreign Policy Concept in 2014 and appeared in the most recent version (that of 2023) as the nation's second-highest priority, just after what Moscow calls the "near abroad" (the post-Soviet space). A naval doctrine published in 2022 insists, too, on the strategic role of the Arctic for Russia's security, and domestic development challenges are well outlined in the 2020 Strategy of Development for the Arctic up to 2035.

Russia's policy of "reconquering" the Arctic, both in terms of its international status and in terms of regional development, is a response to legitimate domestic concerns and the need for Moscow to assert its sovereignty, both symbolically and in practice. Almost 30 percent of the country's land territory lies north of the Arctic Circle (4.8 million square kilometers or about 1.85 million square miles), and even if the Arctic Zone is home to fewer than 3 million people (7 million if we add in the sub-Arctic territories), it generates between 10 and 15 percent of Russia's GDP and a quarter of its exports. In 2020, 80 percent of natural gas and 17 percent of the country's oil were produced in the Arctic regions, and the yet-to-be-explored reserves of the continental shelf are immense. In an age of climate change, this high-level dependence on fossil fuels and their Arctic location could be a sword of Damocles hanging over the head of the Russian state.[1] Therefore, the importance of this region for Moscow is substantial in the long term, regardless of internal political developments and the evolution of its relations with the West.

However, this importance is hampered by obstacles specific to the Russian political and economic system, as well as by the realities of the Arctic, including climatic conditions and isolation. A legacy of its Soviet heritage, Russia is indeed the only Arctic state to have developed an extensive human and industrial presence in such inhospitable areas. This utilitarian vision of the region, seen as a resource to be exploited, involves human and financial investments that the Russian government cannot fully sustain under present social and budgetary conditions. Russian Arctic development policy must therefore take into account the economic demands of the national budget and the geopolitical realities of decoupling from the West and pivoting toward the so-called Global South. Additionally, it must consider the heavy legacies of the past in terms of population, infrastructure, and pollution, as well as future projections regarding the impacts of climate change.

This book drives us through Russia's Arctic policies by looking first at the strategic stakes at play in the Arctic region in terms of boundary-making, climate change, and investments in the NSR. Next, it explores the Arctic as a diplomatic platform for Russia's interactions with other Arctic players, both Western countries and those in the Global South, as well as the dual nature of the current remilitarization of the Far North. It then moves on to the economic challenges to be addressed by Russia in its quest for energy and mineral exploitation of the huge Northern Siberian landmass and its adjacent maritime spaces, especially at a time of massive Western sanctions. Last but not least, it delves into the development issues of Russia's Arctic cities, shaped by their Soviet legacy, demographic and sociological transformations, and the difficulties of Indigenous peoples to find a room for themselves in Russia's ambitious plan for the Arctic.

CHAPTER 1
A STRATEGIC FACADE SUBJECT TO INCREASED ATTENTION

For Russia, the Arctic has long been a unique and complex strategic space whose importance has only continued to grow. For what use were these 19,724 kilometers (12,255 miles) of mostly uninhabited coastline, these maritime expanses bordered by ice floes, and their challenging navigation conditions? For a long time, the Northeast Passage was nothing but a myth, a dream for explorers and adventurers. However, for Russia this myth gained a practical significance very early on, during the reign of Ivan IV (known as "The Terrible") in the sixteenth century. English merchants showed the tsar that it was possible to reach Northern Europe by rounding the Kola Peninsula, whose northern shores are ice-free all year round. Arkhangelsk, founded in 1584, became the first international port of the emerging Russian Empire. Expeditions and explorations followed one another for four centuries, and the importance of these regions continued to grow. Beyond the perpetual quest for the Northeast Passage, which for the Russians became known as the Northern Sea Route (Severnyi morskoi put' or Sevmorput'), the development of numerous mineral deposits, such as gold, diamonds, and rare metals, and later hydrocarbons, contributed to the transformation of the Arctic into an energy and mineral bonanza.

The Stakes of a Space with Contested Boundaries

It is not easy to define the perimeter of Russia's Arctic space, which closely intertwines maritime expanses partly covered by ice floes with a collection of lands, islands, and the northern reaches of the vast Siberian landmass. Several legal issues are at stake: the definition of the continental shelf, the maritime border delimitations with Norway and the United States, the status of the Northeast Passage and the Northern Sea Route, and the definition by the Russian state of its "Arctic Zone," an issue of domestic balance between the Siberian regions and the support they receive from the state (see Map 1.1).[1]

The Continental Shelf Battle

The Russian authorities cared about the border delimitations of their country up to the North Pole quite early in time. As early as 1926, the Soviet government, concerned about the possible discovery of unknown lands by Western countries that had better mastered aviation than the Soviet Union then had, issued a decree "On the Proclamation of Lands and Islands Located in the Northern Arctic Ocean as Territory of the USSR." The decree stated that "all lands and islands, both discovered and which may be discovered in the future, which do not comprise at the time of publication of the present decree the territory of any foreign state recognized by the Government of the Soviet Union and located in the northern Arctic Ocean … are proclaimed to be territory of the Soviet Union."[2] Moscow thus claimed sovereignty over all the territories up to the North Pole between the internationally recognized boundaries of the time: to the east, those between the United States and Russia defined in the 1867 Convention on Alaska; and to the west, the border between the Soviet Union and Norway.

With the collapse of the Soviet Union, and despite media-hyped depictions of a forthcoming "ice war" in the Arctic, none of the five coastal states (Russia, Norway, Denmark, Canada, and the United States) are involved in violent confrontation or unlawful occupation of disputed maritime territories. State behavior is guided by the agreed-upon rules of international law, and territorial disputes have been characterized as much by symbolic competition as by pragmatic cooperation.[3] The five states bordering the Arctic Ocean and its adjacent seas have indeed been attempting to delineate the limits of their territorial waters and the adjoining continental shelf at the United Nations Division for Ocean Affairs and the Law of the Sea (UNCLOS). The stakes are considerable because extending the continental shelf beyond the 200-nautical-mile limit defining territorial waters would offer privileged access to underwater mineral resources (hydrocarbons, rare minerals) that all experts consider significant.[4]

As early as 2001, Russia submitted its own claim arguing that the Lomonosov Ridge and the Alpha-Mendeleev Ridge are both geological extensions of its continental Siberian shelf and, thus, that parts of the central Arctic Ocean, as well as parts of the Barents Sea and the Sea of Okhotsk, fall under its jurisdiction.[5] The Lomonosov Ridge is a 1,800-kilometer-long submerged elevation joining the Eurasian and American continental crust, while the Mendeleev Ridge is a 1,500-kilometer-long elevation between Russia's Wrangel Island and the Canadian Arctic Archipelago. According to Russian legal experts, this claim would allow the extension of control over an expanded continental shelf practically up to the North Pole. The UN Commission rejected this initial request, arguing that the two ends of the area must first be negotiated with the neighboring countries, Norway and the United States. Regarding the polar area, Moscow was required to complete the dossier with bathymetric maps supporting its claim.[6]

Denmark and Norway—which, like Russia (and unlike the United States), have ratified the 1982 UNCLOS Treaty—contested this claim, but Russia scored several points in this legal and scientific battle. In January 2023, the Commission on the Limits of the Continental Shelf accepted the main elements of the Russian request, subject to final negotiations with Denmark and Canada.[7] However, the legal battle is far from over, especially since the United States, although not a signatory to the 1982 Convention, has recently expanded its claims at the expense of Canada.[8] The US State Department has also opened the possibility of disputes with Russia in the Bering Sea beyond the 82nd parallel north,[9] which, of course, generated virulent reactions in Russia as it considers that the demarcation line extends as far as the North Pole.[10]

While demarcating Arctic maritime borders is typically addressed through international legal tools, there is a basis for potential interstate tension in the future. If the Lomonosov and Mendeleev Ridges are not recognized as part of Russia's continental shelf, Moscow, which has invested billions of dollars to collect the necessary scientific information, could toughen its rhetorical stance, making it less likely to respect international law and prompting it to advocate more binding structures for dispute resolution. Indeed, Moscow considers the 2023 provisional decision a victory that will be difficult to overturn.[11] Should Russia receive a favorable decision from the Commission regarding its claim, whether in part or in their entirety, it will achieve a long-term territorial advantage on the continental shelf. Other Arctic states, especially Canada and the United States, would either not be able to challenge it or at least have to be willing to pay a heavy diplomatic and strategic price for disregarding it. It would therefore modify the Arctic global geostrategic balance, as well as the prospects of future economic exploitation, in Russia's favor.

Demarcating Maritime Boundaries with the United States and Norway

Two other Arctic maritime boundaries have been gradually delimited between Russia and the United States and Russia and Norway. In the 1970s, the United States proposed to the Soviet Union that the two countries enter into negotiations over the length of their common maritime border (the longest in the world, being more than 2,500 kilometers, or 1,550 miles, long) in order to settle points of disagreement, since the exclusive economic zones (EEZs) of both countries intersected in the Bering Sea as well as in the Chukchi Sea. A provisional application for a forthcoming agreement entered into force in 1977 so that day-to-day issues could be regulated, particularly with respect to fishing. Both parties finally signed an agreement on July 1, 1990, resulting in the creation of the Baker-Shevardnadze Line, which is a compromise between a median line and a sectoral line.[12]

The United States ratified the treaty in 1991, but the Russian Parliament (the Duma) never did, arguing that it harmed the interests of the Russian state in terms of fishing and potentially also of oil reserves. In this dispute with the United States, the Soviet-Russian jurisdictional position has been weakened by its inconsistency. Indeed, according to the US statement on the Russian claim to UNCLOS (which is not public), it appears that in its submission, Russia refers to the 1990 agreement on the Bering Sea, which in this case means that the country is now bound to the treaty even without having ratified it.[13] In a current time of US-Russia tensions and with a Trump presidency particularly sensitive to symbolic recognition of US great power status, it is likely that this maritime boundary question would not be resolved rapidly.

The Russian-Norwegian territorial conflict over the Barents Sea has been the most complex, yet also the one resolved in the most expedient way. It bears the stamp of its Cold War geopolitical context (for many decades, Norway was the only member of NATO, along with Turkey, to share a border with the Soviet Union), but it also involved important economic questions, and it carries symbolic weight in terms of nation-building for both countries.[14] About 155,000 square kilometers (60,450 square miles) were under dispute, including the overlapping EEZs within this area and the 20,000 square kilometers (7,800 square miles) of overlapping claims farther north in the Arctic Ocean. Since 1980, after the Soviet Union attempted to engage in oil extraction, both Moscow and Oslo have agreed on a moratorium prohibiting oil and gas exploration and geological prospecting in the disputed area, which means that fishing has since taken center stage in the underlying economic debates over the border demarcation.

Despite the impossibility of reaching a legal agreement, both countries quickly decided to cooperate regarding fishing. As early as 1978, an agreement concerning a so-called Grey Zone was signed. The 65,000 square kilometers (25,350 square miles) of the Grey Zone include the Loophole—a high seas triangle bounded by Russia's EEZ, the disputed waters between both countries, and the Svalbard Fisheries Protection Zone—as well as 23,000 square kilometers (8,970 square miles) of Norway's EEZ and 3,000 square kilometers (1,170 square miles) belonging to Russia. The Grey Zone Agreement, extended on a yearly basis, was a classic mechanism of enforcement and control in the management and conservation of fish stocks in international or disputed waters.[15]

Throughout the 1990s and 2000s, regular tensions between the two countries arose over the inspection and boarding of Russian fishing boats by the Norwegian Navy. Yet despite elements of significant tension and a complex geopolitical context, Russian-Norwegian cooperation has been a success in terms of the everyday management of maritime relations.[16] Pragmatic cooperation has made it possible to overcome legal conflicts and reach a definitive agreement, concluded in April 2010 during then-President Dmitri Medvedev's visit to Norway, and ratified by the Russian Duma in March 2011. Norway has withdrawn some of its territorial claims, and Russia has consented to a shift of the 1926 demarcation line to split the 175,000 square kilometers (68,250 square miles) into two almost equal parts defined by eight points.[17] The endpoint is still unknown because of the undefined edge of each party's continental shelves in the Arctic Ocean. Russia was granted EEZ rights in the area to the east of the boundary that lies within 200 nautical miles of the Norwegian mainland but more than 200 miles from Russian territory.

The treaty is also accompanied by agreements on cooperation over fisheries and hydrocarbon activities in cases where oil or gas deposits extend across the delimitation line.[18] The Norwegian-Russian Joint Fisheries Commission will continue its activities, but the agreement effectively supersedes the Grey Zone Agreement of 1978. On the Russian side, this decision was eminently political. It was made against the advice of the jurists in charge of the dossier at the Ministry of Foreign Affairs, who criticized Dmitri Medvedev for making excessive compromises.[19]

The 2010 Russian-Norwegian treaty leaves unresolved another point of contention, namely that of the Svalbard Archipelago and its largest island, Spitsbergen. The archipelago, covering 61,000 square kilometers (23,790 square miles) in the Barents Sea, is the object of a complex legal debate related to the limits of Norwegian sovereignty as defined in the 1920 Svalbard Treaty, ratified by more than forty states—though not by the Soviet Union, which had no international legal recognition at the time. The treaty contains complex clauses stipulating that ships and citizens of contracting parties are permitted to undertake fishing and hunting on an equal basis on the lands and in the territorial waters of the archipelago, and that all signatory states have equal access to conduct economic activities there. The Svalbard mining code is favorable to foreign investors, so that the taxes paid promote the archipelago but not the budget of the Norwegian state.[20]

Russia has challenged the Norwegian reading of the treaty at different levels. It claims that the historical coexistence of Norway and Russia over the archipelago must be given legal precedence. It raises the fact that Norwegian lawmakers have no legislative grounds for invoking the "territorial sea"—a classical institution of contemporary international maritime treaty law—in order to mark off an EEZ around the archipelago or on its shelf.[21] Norwegian sovereignty is thus allegedly limited to the land—not the sea. Russia also criticizes the fact that Oslo applies Norwegian domestic law to the archipelago, which restricts exploitation rights. Thus, the fisheries regime used by Oslo for Svalbard is more restricted in terms of permitted catch than in the EEZ. In addition, according to Moscow, Norway has unilaterally imposed a mining code for the islands' continental shelf that contradicts the Paris Treaty.

The Svalbard Environmental Protection Act, enacted by Oslo in 2001, could call into question the activities of the Russian state-owned mining company Trust Arktikugol, which exploits the coal reserves of the Coles Bay area. Moscow long defended the economic interests of the mining town of Barentsburg and viewed Oslo's environmental rhetoric as mere subterfuge to obstruct Russian activities on the archipelago.[22] But over the years, the coal business has been declining and Barentsburg has attempted to move toward Arctic tourism (see more on this later). Tensions have arisen again since 2022, with Norway accusing Russia of provocative actions in Svalbard, warning that Russian fishing vessels in Norwegian waters are often linked to intelligence gathering and that subsea cables connecting Svalbard to the Norwegian mainland have been severed by Russian vessels.[23]

The Northeast Passage

Another issue relates to the legal status of the Northeast and Northwest Passages. The Northeast Passage skirts the northern Russian coast, thus linking the Atlantic and Pacific and avoiding a detour via the Suez Canal or around the Cape of Good Hope. The Northwest Passage runs from the Bering Strait past the northern Alaskan and Canadian coasts to the Atlantic between Labrador and Greenland, connecting the Atlantic Ocean with the Pacific Ocean without having to go through the Panama Canal or rounding Cape Horn. A third potential sea lane, the so-called Transpolar Sea Route, directly crosses the middle of the Arctic Ocean, connecting Eurasia with North America.[24]

Of these three sea lanes of communication, only the third, a high-latitude one, presents no legal problems, as it crosses mainly international waters, but it may not open up fully for several decades, if ever. The other two routes are topics of more intense debate: depending on the outcomes of legal battles and the fluctuations of the ice floes, all or part of the two passages are subject to the national regulations of the bordering countries, Russia and Canada, or, conversely, depend on international navigation rules.

The melting of the icecap is not proceeding as quickly on the Canadian side as on the Russian side, making Russia the first country to be affected by the prospect of ice-free Arctic navigation. Stretching 5,770 nautical miles from Murmansk to Yokohama (compared with 12,840 via the Suez Canal), the Northeast Passage does shorten the journey between Europe and Asia, but it faces a series of constraints that have long limited its use. There are, for instance, multiple shifting shoals between the Arctic archipelagos; ships using the route must contend with floating ice or remnants of ice floes under harsh weather conditions (fog, freezing temperatures, snowstorms, etc.), and recent history records multiple accidents such as ships being trapped and crushed by ice or blocked for weeks, especially in the eastern part of the route.[25]

Both Canada and Russia view their Arctic passages as having historically belonged to them, and oppose international opinion, in particular that of the United States, which argues that they are international waters open to free navigation.[26] During the decades of the Cold War, these legal ambiguities served to stoke tensions between the Soviet Union and the United States: US submarines not only succeeded in reaching the North Pole (in 1958, the USS *Nautilus* was the first watercraft to reach the geographic North Pole) but also passed through Soviet-controlled Arctic waters and northern straits (achieved by the USS *Blackfin*), and even entered Soviet territorial waters (the USS *Gudgeon* in 1957, close to Vladivostok).

Whatever their legal status, both Passages are open to foreign commercial traffic, but state prerogatives are more significant if they are recognized as national straits. In this latter case, the state has the right to apply "special conditions" in accordance with the extent of ice coverage, and particularly in cases of severe weather conditions. Ships must give advance notice of their transit, apply for guidance, and comply with national laws. In the former case, that of international waters, all ships enjoy the right of transit passage without having to ask for the authorization of any specific body; the littoral states can only enforce fishing and environmental regulations, fiscal and anti-smuggling laws, as well as laws designed to ensure the safety of ships at sea.

Despite the debates surrounding the legal status of the waters being transited, Russian territorial waters are still subject to the right of innocent passage, and the Law of the Sea Convention requires that treatment of foreign vessels be nondiscriminatory. Russia is thus legally unable to charge fees to transit through its Arctic waters, but may establish regulations governing the passage of vessels in ice-covered areas, especially in accordance with environmental protection and safety laws (such as civil liability regulations for damage arising from vessel-source oil pollution). In 2012, the Duma passed a long-awaited "Law on the Northern Sea Route," which stipulates conditions of transit and demands new insurance requirements, under which responsibility for possible environmental damage and pollution is ascribed to ship owners, and which set costly tariffs for assistance and logistical information.[27]

These binding rules have been validated by major international insurance companies, but have been refuted by the United States, which deems that acceptance of such would be tantamount to recognizing Russia's sovereignty beyond its territorial waters. These costly services—icebreaker assistance, sailing master services, radio communication, and hydrographic information—are provided by the Marine Operation Headquarters and the Northern Sea Route Administration, which has been based in Arkhangelsk since 2013. If it is widely recognized that coastal states should not be solely financially responsible for costs associated with transit, it seems that thus far only foreign vessels are paying for it, and that Russian ships are exempt, which in legal terms can be regarded as a discriminatory measure.[28] As we will see further, even if the Northeast Passage still faces legal tensions related to Russia's rights to apply fees for services, the route is used by foreign, mostly Asian, ships that are ready to accommodate Russia's claims, and Western ships are in any case avoiding the route since the 2022 waves of sanctions that followed Russia's invasion of Ukraine.

Defining Russia's Arctic Zone

The terrestrial boundaries of the Russian Arctic space are also the subject of numerous debates, this time all internal to Russia. As early as 1932, within the framework of the first five-year plan, the Soviet authorities defined northern territories where salary advantages and bonuses were granted. The list of these "Extreme North" (*Krainyi sever*) territories has been revised several times over decades and includes, for economic and salary management reasons, many districts in southern Siberia or the Far East (regions around Lake Baikal, Tyva, or Buryatia), which, although they indeed present extreme living and working conditions, are far removed from the Arctic. In April 1989, at the height of perestroika, the USSR Council of Ministers' State Commission on Arctic Affairs defined the country's Arctic Zone thusly: 3.1 million square kilometers of landmass and about 4 million square kilometers of continental shelf.[29]

After the Soviet collapse, the Russian legislature pushed to specify the contours of regions that are part of a more tightly defined "Arctic Zone." In 1999, the Duma's Commission for Arctic Affairs drafted an initial list that served as the basis for defining the first "Foundations of Russian Arctic Policy until 2020," which was signed by President Dmitry Medvedev in 2008. In 2014, a decree specified the list of "terrestrial territories of the Arctic zone of the Russian Federation," restrictively including all territories located between the shores of the Arctic Ocean and the Arctic Circle. The list was revised in 2017 and 2019 to add some districts of Karelia and the northern regions of Sakha-Yakutia.

Today, the Russian Arctic Zone comprises four "subjects of the federation" in their entirety (the Murmansk region and the Nenets, Yamalo-Nenets, and Chukotka autonomous districts) and the northern parts of five other subjects (the Karelia, Komi, and Sakha-Yakutia republics, and the Arkhangelsk and Krasnoyarsk krais). Officially, neither the Khanty-Mansi autonomous district nor the Magadan region, considered more southerly, are included. In 2020, President Putin enacted a "Law on State Support for Entrepreneurial Activity in the Arctic Zone of the Russian Federation,"[30] demonstrating the authorities' major interest in the economic and military development of this strategic area. However, all these legislations also need to take into consideration the evolving material conditions in a region particularly affected by climate change.

Soviet Adaptations to Natural Constraints ...

All types of human activities face severe constraints in the Far North. Very long winters with polar night and often extreme temperatures pose numerous technical challenges for both people and machinery: the coldest ever temperature in the Northern Hemisphere, −67°C (or nearly −89°F), was recorded near Verkhoyansk in Sakha-Yakutia, which lies within Russia's Arctic Zone. Short summers are feared for transforming large areas into seasonal swamps and the accompanying swarming of destructive insects. And even the supposed light at the end of the wintry tunnel, springtime, is dreaded because of the melting of frozen and snow-covered ground (the infamous *rasputitsa*), which significantly hinders all land transportation.

In addition to the climatic constraints, there are a number of specific problems posed by the thawing of permafrost (known as *merzlota* in Russian), the deep-frozen ground—a fragile glacial legacy of the last ice age—which is now being degraded by human activity

Russia's Arctic

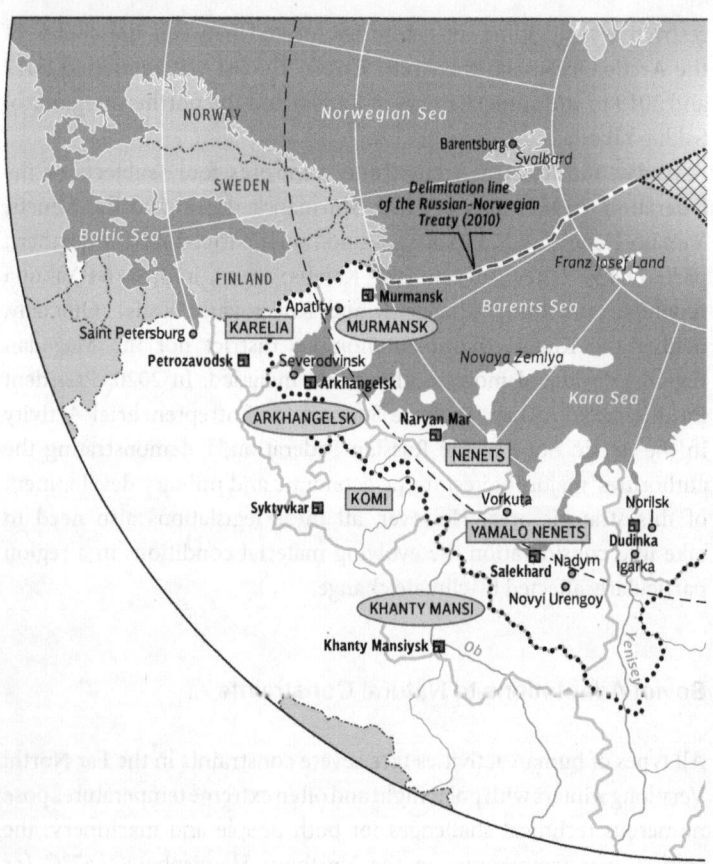

Map 1.1 Borders and territorial claims in the Russian Arctic.

A Strategic Facade

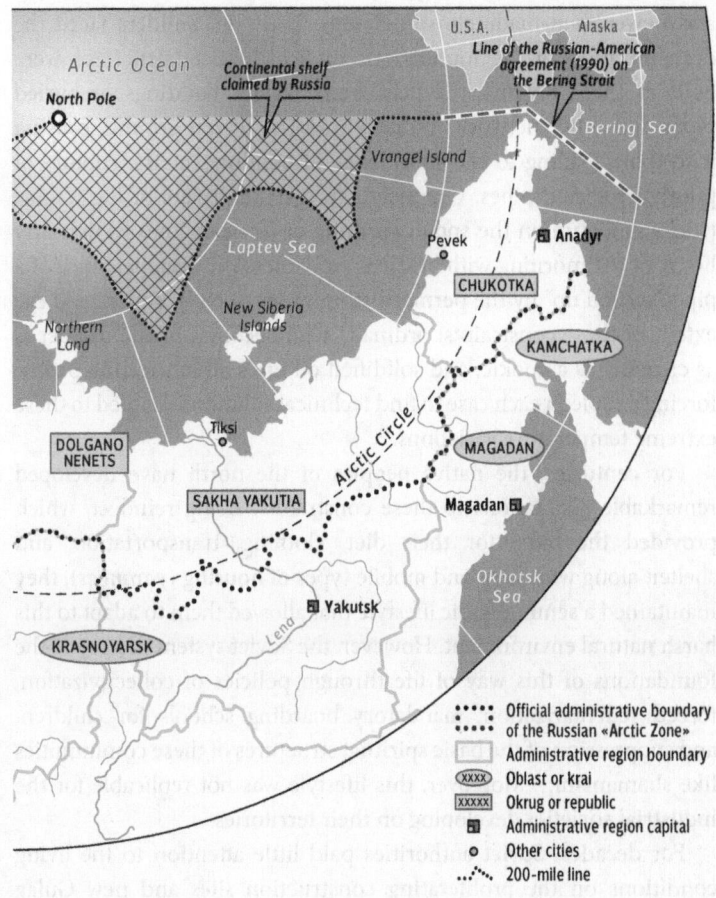

and newer man-made climate change. Early on, builders faced the destabilization of the foundations on which their structures were built in these regions: the heat from surface buildings or buried pipes and other networks radiates into the frozen ground, causing it to thaw, leading to the subsidence of buildings or the rupture of poorly protected pipes. One needs to see the drivers of big Kamaz trucks, lined up on the snow, warming up their engines in the early hours of the morning with braziers, or witness the disappearance of a pipe "sucked up" by the permafrost, in order to truly understand the extent of these constraints: ordinary steel becomes brittle, rubber is as crumbly as a cookie, and solidified oil loses all lubricating power, forcing people in each case to find technical solutions adapted to these extreme temperature variations.[31]

For centuries, the native peoples of the north have developed remarkable adaptations to these constraints. Using reindeer, which provided the basis for their diet, clothing, transportation, and shelter, along with light and mobile types of housing (*yarangas*), they maintained a seminomadic lifestyle that allowed them to adapt to this harsh natural environment. However, the Soviet system destroyed the foundations of this way of life through policies of collectivization, forced sedentarization, mandatory boarding schools for children, and suppression of the basic spiritual structures of these communities like shamanism.[32] Moreover, this lifestyle was not replicable for the industrial societies developing on their territories.

For decades, Soviet authorities paid little attention to the living conditions on the proliferating construction sites and new Gulag towns.[33] Starting with the foundation of the first camp in 1923 on the Solovki Islands in the White Sea, the Gulag system (a Russian acronym for State Camp Administration, the huge Soviet system of penal camps) took charge of most of the Arctic and sub-Arctic deposits being developed. This continued with the Pechora coal basin in 1931, the same year as the creation of Dalstroy ("Far North Construction"), which was responsible for road construction and the development of gold mines in Kolyma (in the Magadan region), and in 1935, the beginning of the exploitation of nickel and rare metal deposits in Norilsk, to name the most symbolic projects. Varlam

Shalamov and Alexander Solzhenitsyn have described the living and working conditions in these camps, where there was no concern for the longevity of a workforce that was exploited to the utmost.[34]

It was only with the end of the Stalinist regime in the mid-1950s that the authorities began to consider other ways to develop these inhospitable lands. Since they could no longer rely heavily on this servile workforce, they had to find other means to attract workers. A complex system of bonuses and various benefits (*orgnabor*) was then implemented to bring in workers and engineers to the construction sites and to populate the towns, both small and large, that were emerging in the Far North. At the same time, medical studies were published showing the health disorders related to long stays at these latitudes.[35]

In fact, depending on the types of construction sites and the specificities of each concrete region, different modes of settlement were developed. On the technical side, Soviet engineers developed increasingly sophisticated methods of construction on permafrost to avoid thawing problems. Beginning in the 1950s, new buildings were systematically built on stilts anchored deep in the frozen ground called "pile foundations." "Such a foundation puts a layer of air between the ground and the building, effectively decoupling the heat generated by the structure from the frozen ground and thus preventing the warming of ice-rich permafrost," explains Nikolai Shiklomanov of the George Washington University.[36] Isolation systems were developed for all buried networks, but most pipes were constructed above the surface, significantly altering these new industrial and urban landscapes.

Moreover, Soviet urban planners gradually devised a wide variety of installations adapted to different regions and their particularities, or in response to the Soviet regime's desire to demonstrate its ability to overcome natural constraints. This was attempted in Norilsk where, after the initial phase of development integrated into the Gulag, planners proposed a sort of ideal Arctic city by creating a real town for a "normal" population (including women and children), which required the creation of a full range of services despite its location north of the Arctic Circle.[37]

Norilsk and other Arctic cities began growing fast with prefabricated concrete building elements, which could be quickly assembled on a pile foundation to construct large multistory housing. "As a result, the rate of construction of new residential buildings in Norilsk increased from five per year in the 1950s to approximately 18–20 per year from the 1960s to the late 1980s. Construction on piles was considered to be a major engineering achievement, prompting the Soviet media to proclaim that the 'Permafrost is conquered,'" Shiklomanov adds.[38] At the end of the Soviet Union, more than 75 percent of structures in Russian permafrost regions were constructed on pile foundations.

This very particular experience, long praised by the Soviet authorities, faced sharp criticism in the 1980s for its cost, enormity, and the ecological disorders it caused. At the same time, entirely different approaches favored lighter or temporary modes of settlement that took into account both the specific environment and the lifespan of the deposits, limited by the volume of their reserves. Thus, various models of "shift cities" (*vakhtovyi gorodok*) appeared, consisting of small installations where teams of specialized workers, much like offshore drilling platforms, took turns. Their permanent homes and families could be found in southern Siberia or even in the European part of the country. These shift cities could simply consist of a group of thermally insulated containers serving as sleeping quarters, canteens, or infirmaries, set up and moved as needed by helicopters to various construction sites.

… Challenged by Climate Change

However, as more northern construction sites proliferated, these different types of adaptations were challenged by the effects of climate change. In the Arctic Ocean and its adjacent seas, warming has caused a dramatic retreat of the ice cover, with significant annual variations. The extent of summer ice has decreased from an average of about 7 million square kilometers (2.7 million square miles) in the 1980s to a little over 4 million square kilometers (1.5 million square miles) in 2024—near historic lows already experienced in 2020 and 2021.[39] For the Northern Sea Route, off the Russian coast, this means both easier passage conditions and an extended navigation season, which is not perceived without apprehension by the authorities. Indeed, if the retreat of the ice cover continues to accelerate, there may come a time when ships of all nationalities can pass completely outside Russian territorial waters, through the Central Lane of the Arctic Ocean, thus escaping the control that Moscow intends to maintain over this route.

On the continent, the effects are considerable and have been the subject of numerous studies in Russia, Scandinavia, and Canada. There is a significant expansion of wetlands, swamps, and peatlands in the summer, and the gradual replacement of plant formations. The tundra is shrinking, replaced by grasslands invaded by more southern vegetation, while the composition of boreal forests is being altered.[40] The thawing of permafrost (which can reach several hundred meters deep in the large sedimentary basins of the Ob, Yenisei, and Lena rivers) also releases significant amounts of carbon dioxide and other greenhouse gases, including methane, which only further accelerate global warming.[41]

The US National Oceanic and Atmospheric Administration (NOAA) has noted than 2024 was the first year when the tundra changed its natural regime and began to emit more carbon dioxide than it absorbs due to higher temperature and more regular large-scale fires.[42] This mass thawing could gradually transform certain Arctic regions into a mosaic of land and water, thus worsening problems of connectivity and the state of transport networks.[43] This phenomenon is exacerbated by the increasing frequency of forest and tundra fires. Concerns, not yet fully verified, have been raised about the potential reactivation of various ancient pathogens, viruses, and microbes trapped in this ice.[44]

The authorities of Arctic states are attempting to calculate the foreseeable effects of this thawing on the future of various kinds of infrastructure, residential buildings, and other structures that have been built with varying degrees of caution in these regions. Russia, where urban and industrial installations have proliferated the most, is certainly the most affected in this regard. The permafrost thaw's effects are indeed often spectacular: collapsed buildings, and roads damaged communications, and ruptured gas pipelines multiply. In

Norilsk, where these effects are particularly dramatic and undoubtedly exacerbated by the haphazard exploitation of the subsoil, around 60 percent of the buildings have already been deformed by the thawing ground.[45]

Already in the 1970s, Arctic engineers noticed the collapses of concrete buildings in Norilsk and Yakutsk, attributed to the reduced bearing capacity of pile foundations due to permafrost warming. This trend only accelerated in the 1990s and 2000s, with a growing part of Norilsk's urban infrastructure at risk of collapse. Climate change is not the only culprit: Arctic cities are also altered by the urban heat island effect, when urban temperatures rise by several degrees relative to their surrounding environment because of the use of concrete, and therefore of sand, and different mechanical and thermal pressures put by infrastructure on the frozen ground.[46] A recent study has calculated that around 20 percent of all industrial and transport infrastructure in Russia's Arctic, and more than 50 percent of its residential buildings, will be affected by permafrost thawing by 2050, at an estimated cost of $250 billion.[47]

... And by Russia's Climate Policies

Yet Russia's position on climate change remains ambivalent. The Russian scientific community has studied the evolution of the polar climate for several decades, and during Soviet times interpreted changes as natural variations of the climate. Since then, Russian experts have been divided between those who attribute climate change to mainly anthropogenic factors and those who continue to prefer the idea of a natural cyclical evolution (the "Earth's cycles," in the words of Vladimir Putin).[48]

Politically, the regime is playing on both views, opportunistically, depending on the audience and the situation. Three main lines of reasoning put forward by Moscow can be identified:[49] (a) climate change is real but non-anthropogenic and is part of a Western campaign against Russia's reassertion on the international scene; (b) climate change is real and anthropogenic, but it will bring mostly positive changes for the country (development of agriculture in northern regions, easier navigation and access to new deposits of raw materials, etc.); and (c) climate change is real, anthropogenic, and negative, but Russia will nonetheless continue to give priority to its energy and extraction policies, because the country cannot afford to develop alternative economic strategies and will limit the impact of change through adaptation measures.[50]

Moscow long hesitated to ratify the Kyoto Protocol to reduce greenhouse gas emissions, launched in 1997 at the initiative of Japan and the European Union. Many Russian economists and experts opposed it, arguing that the protocol's effects would hinder the recovery of the national economy after the severe crisis of the early 1990s that followed the collapse of the Soviet Union. Vladimir Putin did not seem inclined to ratify it, declaring in 2003, "We often hear, jokingly or seriously, that for a northern country like Russia, a 2 to 3 degree warming wouldn't be a problem and might even be beneficial. We would spend less on fur coats and warm clothing. And agronomists tell us that agricultural production could increase."[51]

It was therefore a surprise to everyone when the Russian president announced in 2004 that Moscow would ratify the treaty.[52] Russia's signing of the treaty, as a state representing 17 percent of global greenhouse gas emissions at the time, enabled the protocol to reach the 55 percent of global emitters necessary for its implementation, despite the United States' refusal to sign. Debates continued in Russia, but Vladimir Putin defended his decision and held numerous interministerial meetings on the topic. In 2019, at the Arctic Forum in Saint Petersburg, he indicated that, according to Russian sources, the Arctic is warming four times faster than the global average, and northern Russia 2.5 times faster.[53] In 2021, at the World Economic Forum, he discussed the threats to northern Russian cities: "Entire cities in our Arctic region are built on permafrost. If everything starts to melt, imagine the consequences for Russia. Of course, we are worried."[54]

These speeches cannot hide one of Russia's critical points of ambivalence: if the country's greatest legacy to the planet is its role in biodiversity and wilderness preservation, it has failed to advance effective environmental legislation and has gradually lowered its standards of environmental protection.[55] Several Russian institutions such as the Academy of Sciences and the Institute for Economic Forecasting have studied how much growth the Russian economy could experience from global warming, mostly thanks to an expansion of agriculture in the north and to the use of the Northern Sea Route.[56] Yet such an optimistic forecast does not take into consideration the destruction of infrastructure as a result of the thawing of the permafrost in the north, and more frequent flooding and drought in the southern parts of the country.

The Russian authorities do not deny the environmental consequences of the country's industrial and military activities in the region over the decades. For example, Russian researchers have identified twenty-seven areas in the Russian Arctic affected by pollution to the point of causing severe environmental damage and increased mortality among the population, such as the Murmansk region, the surroundings of Norilsk, and regions in western Siberia with large investments in oil and gas exploration.[57]

For several years, the Russian government has taken measures to clean up certain polluted areas, often in the context of joint projects with the Arctic Council and the Barents Euro-Arctic Council. These include cleaning up metallic waste left behind by military infrastructure on the archipelago of Franz Joseph Land and on Wrangel Island, the dismantling of hundreds of rusty ships, the decontamination of certain Soviet nuclear submarines stationed on the Kola Peninsula and the safe disposal and storage of their nuclear waste, and so on. However, some other projects have not been completed, such as the cleaning up of the mining towns of Svalbard, and regular incidents confirm that the safety risks are numerous and often poorly managed.

Industrial pollution is difficult to tackle because it directly contradicts the government's economic objectives. In 2020, the spill of thousands of tons of diesel in Norilsk caused a global media frenzy and prompted a firm response from the Russian president, declaring a state of emergency and criticizing the holding Nornickel's (Norilsknickel) lack of response and failing to report it in time.[58] However, such disasters, produced by a combination of negligence, lack of foresight, and the effects of warming, remain a constant risk in a particularly fragile natural environment that takes decades to recover from any human intervention.

Urban life is also deeply affected by these environmental issues. Eight of the world's twelve Arctic cities with more than 100,000 inhabitants are in Russia. These cities face two parallel phenomena: the "greening" and the "browning" of land. The first, occurring in the tundra, describes the lengthening of the seasons favorable to the growth of vegetation and the appearance of more southern flora, due mainly to the rise in local temperatures linked to industrial production. It is estimated that the bioclimatic zones of Siberia will move north by 600 kilometers by the end of the century.[59] This greening contributes to the arrival of new fauna, especially insects, which increases the risk of pandemics, while also opening up new agricultural opportunities.[60] As for browning, it is occurring in more southern areas, those of the taiga, around industrial cities (generally within a radius of five to ten kilometers, and sometimes beyond, such as with the technogenic deserts of Norilsk and Nikel). Such browning is accompanied by a decline in land output due to pollution linked to extraction activities and multiple chemical contaminations, not only from industry but also from transport systems and urban activities such as heating.[61]

Against this bleak backdrop, a notable exception has been the expansion of the protected area system. A new law on the status of protected habitats and nature reserves was enacted in 2022, with the creation of several new national parks in Kamchatka, Krasnoyarsk Krai, and the Yamalo-Nenets Autonomous District. But this appears to be an easy way to greenwash national policy by protecting some territories from human exploitation, without any compulsion to address energy inefficiency and pollution in the country's industries. Concerns related to permafrost thawing have also resulted in the first "Law on Perennial (Eternal) Permafrost," which came into effect in the Yamalo-Nenets District on January 1, 2024. This law imposes various protective measures and provides aid to local populations.[62] In addition to significant impacts on the economic development of the entire area, climate change also implies new constraints for Arctic operational modalities.

Russia's Arctic

New Investments for a Strategic and Coveted Corridor

How does one reach the Russian Arctic? One first thinks of the Northeast Passage, which long attracted explorers seeking to travel from the Atlantic to the Pacific or reach the Pole, either by free navigation or by drifting with the movements of the ice floes. But other explorers sought to map the lands of the Far North, where mountaineers and geologists discovered mountain ranges as late as the 1920s (such as the Chersky Range, identified in northern Yakutia in 1926).

With the discovery of numerous kinds of natural resource deposits and their exploitation starting in the 1930s, the question of their accessibility—either by sea from the north or by land and river from the south—arose repeatedly. These deposits are often isolated, separated by hundreds of kilometers from their nearest human settlement, and their locations rarely allow access by sea. Transportation is therefore a fundamental element of their exploitation cost, with notable differences from one commodity to another. Gold from Magadan or Sakha diamonds can be transported by plane or helicopter. However, coking coal from Pechora or nickel from Norilsk, and more generally all metals with low-grade ores that must be enriched or refined on site, require completely different industrial and transportation logistics.

In addition to these economic and technical issues, another problem quickly emerged. Early on, the Northern Sea Route and its adjacent seas became a major strategic concern for the Russian government: its protection and military use became unavoidable components of all development projects in this region, which was quickly classified as a border zone with restricted access conditions, inaccessible to foreigners or even the average Soviet citizen without special authorization.

As the mineral wealth of these regions began to be understood, the central role of the railway, the only efficient year-round mode of land transportation, became evident. Arkhangelsk received its first railway connection in 1898, and Murmansk followed in 1916. After the October Revolution, Soviet authorities were hoping to replicate for Northern Siberia the role the Trans-Siberian Railway played for

Southern Siberia's development. In 1928, during the first Soviet five-year plan, the government envisioned the construction of a "Northern Siberian Railway" (Severosibirskaia zheleznodorozhnaia magistral') that would serve the coal deposits of Pechora, the northern Urals, the middle Ob region, and extend to Chukotka or the Strait of Tartary (separating the Pacific Island of Sakhalin from the Russian mainland).

Gulag prisoners built the Vorkuta line between 1937 and 1941, but the Second World War halted this grand project. Only a few segments were completed, such as the Vorkuta to Labytnangi line between 1947 and 1953, and the line further east connecting Salekhard to Nadym and Urengoy, but the planned bridge over the Ob was never built. Similarly, the project for a line to Yakutsk, considered as early as the 1930s, only saw the beginnings of construction after 1975, and the capital of Sakha-Yakutia still does not have train service to this day. The considerable width of the Ob and Lena rivers, which is at their widest when the snow melts, makes the construction of Arctic and sub-Arctic bridges particularly costly. Far from being abandoned, this project recently resurfaced as one of the priority measures announced by Putin in his speech at the Murmansk Arctic Forum in 2025.

Thus, even today, the exploitation of various natural resource deposits relies on multiple modes of transportation: rail lines, where available; roads, with their seasonal constraints; and river segments subject to their navigable season and flow variations (e.g., low water at the end of summer often requires postponing shipments on the Yenisei north of Krasnoyarsk because ships cannot pass the nearby rapids). These constraints were taken into account when developing the gas fields in the northern Ob Plateau, considering that pipelines became the primary means of transporting hydrocarbons both domestically and internationally from the 1960s onward.

Reviving the Northern Sea Route

At the same time, the Northern Sea Route has received continuous attention by the Russian authorities due to growing economic motivations, alongside strategic and military interests. The first recognized passage from west to east along the Northern Sea Route in a single navigation season dates back to 1932 with Otto Schmidt's expedition, and the first commercial passage to 1935. For decades, the route mainly served limited coastal navigation in a few sectors, primarily serving Dudinka (the river port of Norilsk on the Yenisei) from Murmansk or Arkhangelsk and a few ports and bases in the Far North from Vladivostok. The arrival of nuclear-powered ships (the *Lenin* launched in 1960 and the *Arktika* in 1972) extended the navigable season, open year-round from 1978 between Murmansk and Dudinka and from June to October between Dikson and Vladivostok. However, exceptional years must be accounted for: in 1983, several dozen ships were blocked, and a cargo ship was sunk by ice off Pevek.[63] Thus, despite the promises and appeals of Soviet and then Russian authorities to attract potential customers, traffic has remained modest.

Although Russia's Arctic coastline stretches more than 14,000 kilometers (8,700 miles) across the Barents, White, Kara, Laptev, and East Siberian Seas, the Northern Sea Route proper does not include the Barents Sea and is considered to lie between the port of Kara, at the western entrance of the Novaya Zemlya straits, and Provideniya Bay, at the southern opening of the Bering Strait, which makes a total length of 5,600 kilometers (3,480 miles). It covers nearly sixty straits, the main ones being the Vilkitskii, Shokalski, Dmitri Laptev, and Sannikov Straits, running through three archipelagos: Novaya Zemlya, Severnaya Zemlya, and the New Siberian Islands.[64] The legal definition is thus particularly complex, as there is not one single shipping channel per se; rather, there are multiple lanes, and the Northern Sea Route passes through waters of different status: internal, territorial, and adjacent waters, as well as both Russia's EEZ and the high seas.

After peaking at 6.4 million tons in 1986, traffic collapsed to below 2 million tons in the 1990s. Amid the post-Soviet economic crisis, the Russian state could no longer maintain its icebreaker fleet, while private interests grew keener. In late 2002, Yukos Oil Company owner Mikhail Khodorkovsky, along with Lukoil and Sibneft, proposed the creation of a private deepwater oil port in Murmansk.[65] Connected by private pipelines to the Ob oil fields, it would have allowed the sale of oil via large tankers to the United States. Just as much as Khodorkovsky's political opposition to Putin, this project for the total autonomy of Yukos from state oil export structures was a major factor in the oligarch's arrest, and the project was halted. Traffic on the route stagnated for a long time at around 4 million tons, consisting mainly of timber and enriched metals from Norilsk.

It was only with the start of gas field exploitation in Yamal and the construction projects for Novatek's gas (see Chapter 3) liquefaction plants that traffic really began to increase, aided by climate change. Until the late 2010s, pipelines remained the preferred mode of export, as evidenced by the Yamal-Europe projects and the desire to connect the northern Ob Plateau to the southern Siberian pipeline network toward the Pacific, particularly through the Eastern Siberia Pacific Ocean (ESPO) pipeline from Tayshet to Kozmino, announced as early as 2001 and inaugurated in 2012; and the launch in the same year of the Power of Siberia (Sila Sibiri) Gas Pipeline linking Sakha to Vladivostok, planned as early as 1997 by President Boris Yeltsin and inaugurated in 2019 under Putin.

From 4 million tons in 2014, the total traffic of the Northern Sea Route increased to 7.5 million tons in 2016, 36 million tons in 2023, and 37.3 in 2024 (see Graph 1.1).[66] Yet this impressive growth remains well below Putin's wishes of 80 million tons, and the official forecasts, which had projected 90 million tons by 2024, 150 million tons by 2030, and 220 million tons by 2035, seem difficult to achieve given the current conditions in terms of both Western sanctions and world commodities market.[67] The Russian route is thus far from competing with the Suez Canal—the latter saw 23,000 ships transporting 1.4 billion tons in 2022. It is also worth noting that the majority of current traffic on the Northern Sea Route is still related to Russian coastal shipping (i.e., from one Russian port to another). Actual transit traffic (i.e., ships connecting a European port to a non-Russian Asian port) represented only about 2.1 million tons in 2023, or approximately 6 percent of the total.[68] But transit voyages through were set to hit record numbers in 2024, with most of the shipments moving between Russia and China, and carrying 2.38 million tons of crude oil, iron ore, coal, fertilizers, and now more and more liquefied natural gas.[69]

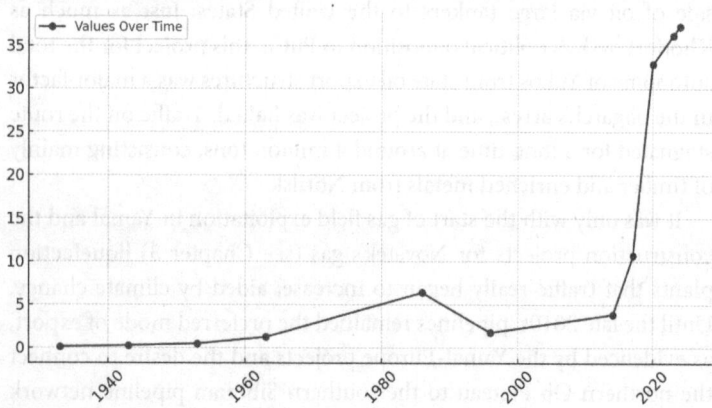

Graph 1.1 Total traffic of the Northern Sea Route (millions of tons, coastal and transit).

Source: M. Kachka, *Atomnyi ledokol'nyi flot—kliuchevoe zveno v razvitie Severnogo morskogo puti* (Moscow: Rosatom, Rosatomflot, 2017), and Russian media.

The Russian government has been sending mixed signals to foreign actors interested in the Northern Sea Route. It has imposed a series of rules for passage, which has been open to foreign ships since 1991: passage in convoys preceded by one or two icebreakers, the obligation to pay for a Russian pilot and specific insurance, and payment of high fees intended, among other things, to finance a set of technical aids, including a weather network, the Russian GPS network, helicopter iceberg detection, and so on.[70] In 2018, it made the state nuclear corporation Rosatom, which already managed all Russian nuclear development, both civilian and military, the official operator for all Northern Sea Route infrastructure.[71] It also prohibits foreign ships from transporting oil, gas, and coal, with one notable exception for the Russian firm Novatek, whose icebreaking LNG carriers fly foreign flags.[72]

This administrative tightening has been accompanied by new legislation, passed in March 2019, which requires foreign warships to notify the Russian government of their passage by way of the Northern Sea Route, forty-five days in advance.[73] According to international maritime law, only passage within the twelve nautical miles of a nation's coastline, comprising its territorial waters, requires authorization, unlike passage on the high seas. But for decades, Moscow has been pursuing an assertive policy of "nationalizing" the entire Northern Sea Route, treating it as though it were an integral part of its national territory. The contradictory messages sent by Moscow to its foreign partners about the conditions of use of the Northern Sea Route therefore hamper Russia's ambitions for a waterway whose infrastructure would be largely financed by foreign capital. Even Russia's faithful Chinese partner has complained about changing standards and Russian dependence on external funding for the route's infrastructure.

Western boycotts of Russian hydrocarbons, and sanctions related to the Russia-Ukraine War, have obviously impacted the expansion of the Northern Sea Route. Moscow had to launch a new development plan in August 2022, including 1.8 trillion rubles for dealing with the disappearance of Western partners.[74] In 2023, Rosatom and Novatek promised year-round navigation in the eastern sector of the route for 2024, thus increasing gas deliveries to Russia's Asian customers, who are expected to absorb 80 percent of the deliveries. The transit time from the Arctic LNG 2 plant to Murmansk is estimated at three to four days, and fifteen to twenty-four days to the East for supplying various Asian customers.[75] However, the new sanctions decided on in June 2024 aim to stop or slow down these LNG flows (see Chapter 3).

But Russia continues to make plans for the Northern Sea Route. It has, for instance, launched the project of a new "ice navigator," planned to be in operation in 2026 and able to offer very precise navigation support for all ships on the Northern Sea Route thanks to advanced satellite imagery called New Cosmos, sponsored partly by Roskosmos, the state space agency.[76] The "Polar Express," a project of a 12,650-kilometer-long submarine communication cable connecting Murmansk and Vladivostok along the Northern Sea Route, launched in 2020 and planned for completion around 2026, is another example of the undoubted will of the Russian government to invest over the long term on Arctic trade, including digital circulation.[77] In addition to the necessity of exporting raw materials from the Arctic Zone, a crucial element of the national budget, the desire to increase Russia's military presence on this strategic front, also plays a major role in Moscow's commitment to the development of the Arctic.

CHAPTER 2
A MAJOR DIPLOMATIC AND SECURITY SPACE FOR MOSCOW

During the Cold War decades, the Arctic was one of the main strategic fronts for the Soviet Union. Military, in particular nuclear, arsenals were stationed there, in close proximity to Europe but also to the United States, which could be quickly reached by missiles flying over the North Pole. In the 1990s and 2000s, the Arctic became a region of dialogue between former adversaries, where multiple diplomatic, economic, and cultural partnerships could be tested.[1] However, the hope of keeping the region free from geopolitical tensions related to Ukraine has failed: the Western Arctic, where the main economic activities and strategic assets are located, is intrinsically linked to the Baltic Sea region, one of the key points of tension between NATO and Russia on their shared neighborhood. Gradually, the Arctic has become entangled in the general geopolitical crisis affecting Western-Russian relations, leading to a remilitarization of the region, putting a halt, or at least a brake, on regional cooperation and accelerating the arrival of Asian countries in the Arctic landscape, symbolizing Moscow's pivot to Asia.

A Space for Diplomatic Socialization

On October 1, 1987, the general secretary of the USSR, Mikhail Gorbachev, delivered his Murmansk Speech, a text that announced the perestroika of Soviet foreign policy, proposing to transform the Arctic into a "zone of peace": denuclearization of Northern Europe, reduction of military activities in the Baltic Sea–North Sea–Greenland Sea area, economic and environmental cooperation projects, opening of the Northern Sea Route, and the creation of collective Arctic institutions.[2] Even though not all the goals of the Murmansk Speech were achieved, particularly those pertaining to denuclearization, the following years saw a profound transformation of the polar geopolitical environment.

In the context of perestroika, exchanges and meetings were organized at all levels between the Soviets and Americans. In the Arctic, direct flights between Anchorage, Fairbanks, or Nome in Alaska, and Vladivostok, Magadan, Provideniya, or Khabarovsk on the Soviet side multiplied. In September 1990, Republican Walter J. Hickel, who was to become Alaska's governor, launched the idea of a Northern Forum that would bring together circumpolar state and regional leaders. Founded in 1991, it received the active support of Mikhail Nikolayev, who was elected president of Yakutia the same year, and this citizens' forum grew to twenty-five regions from the eight Arctic countries in 2001.[3]

The new geopolitical context was embodied in the multiplication of new institutions: the creation of the International Arctic Science Committee in 1990, the International Arctic Social Sciences Association in 1991, and the Arctic Environmental Protection Strategy in the same year, which led to the Arctic Council being inaugurated in 1996. Over the years, these institutions helped to overcome the divides inherited from the Cold War and to integrate Russian Arctic actors into more global frameworks. The Arctic Council has become the main regional institution uniting the eight states with territories lying on or north of the Arctic Circle: Russia, the United States, Canada, Finland, Norway, Sweden, Iceland, and Denmark (through its Greenland territory), with a core of five states with access to the Arctic Ocean proper (Russia, the United States, Canada, Norway, and Denmark). It also grants permanent member status to indigenous nations, although this status is only consultative.[4]

The Arctic Council hosts several working groups that form the core of its daily operations, addressing topics ranging from climate change and environmental monitoring to public health issues and sustainable development.[5] The contributions of Russian researchers, particularly climatologists, oceanographers, and permafrost specialists who have access to data covering more than half of the Arctic lands, are being crucial for polar science and for climate change science more broadly. Indigenous peoples have also been able to cooperate with their Russian counterparts to open a truly circumpolar perspective on the challenges they face.

As demonstrated by one of Russia's leading Arctic researchers, Alexander Sergunin, Russia has long viewed the Arctic Council as a representative multilateral institution—though Western countries remain dominant on it—and has heavily invested in pan-Arctic agendas.[6] Moscow has indeed significantly contributed to developing the procedural rules and mandates of the Council. It has collaborated numerous times with the United States on issues of environmental protection and scientific cooperation, as well as on more sensitive topics such as the American proposal for an Arctic Coast Guard Forum in 2015. Until 2014, Russian authorities were in favor of transforming the Arctic Council from an intergovernmental discussion forum into a full-fledged international organization that would address security issues, including military matters. However, Moscow gradually downplayed this possibility as tensions with the West increased, refocusing its vision of the Council on socioeconomic and environmental issues and on soft, rather than hard, security concerns.

Since the end of the Cold War, Russia has interpreted the Arctic as a space for dialogue rather than conflict, seeking to position itself as a constructive power that respects the status quo of the international liberal order and adheres strictly to international treaties, particularly on issues of seabed delimitation.[7] Some topics, however, have remained more problematic than others. For example, Russia has been criticized by its European partners for its lack of commitment to climate change issues and reducing the carbon footprint of its energy sector, as well as for its treatment of Indigenous populations (see Chapter 4).[8] But overall, Arctic partnerships have developed harmoniously, and in 2011, the Arctic Council was able to announce the creation of a major agreement, the Search and Rescue (SAR) Agreement, which divides the Arctic space into zones of responsibility for each coastal country. This agreement was made possible thanks to the commitment of the

Russian Ministry of Emergency Situations, then headed by Sergei Shoigu.[9]

Many researchers have developed the notion of "Arctic exceptionalism" to describe the maintenance of Russia's constructive position in the region even as Moscow challenged the liberal order on the international stage.[10] As Valery Konyshev and Alexander Sergunin explain, "Moscow does not pursue a revisionist policy in the Arctic. On the contrary, Russia is a status quo power that wants to resolve regional disputes by peaceful means, with the help of international laws and organizations."[11] Indeed, all the conditions were in place for a positive regional dynamic: no direct territorial disputes between Arctic states; clear international treaties on issues related to the Law of the Sea; an Arctic Council that, from its inception, explicitly excluded strategic and military issues from its agenda; the United States and NATO being careful not to geopoliticize the region; Scandinavian countries interested in developing local cross-border relations; and a Russia seeking recognition and prestige.[12]

This prestige-oriented strategy was complicated by Russia's nation-building strategies, embodied by the president's special representative for cooperation in the Arctic and Antarctic, the famous polar explorer Arthur Chilingarov, a member of Putin's United Russia Party and close associate of Putin. During the World Meteorological Organization-sponsored Polar Year in 2007, he organized a helicopter flight to the South Pole and the Amundsen-Scott station in the company of Nikolai Patrushev, then director of the FSB, and led a highly publicized Russian expedition to the North Pole.[13] The nuclear icebreaker *Rossiia* and research ship *Akademik Fedorov* reached the North Pole, where two deepwater submersibles, the *Mir-1* and *Mir-2*, were launched to plant a Russian flag on the Arctic seabed, at a depth of about 4,300 meters (14,100 feet).[14] Chilingarov stated that "we have exercised the maritime right of the first night,"[15] while in 2009, he again bluntly asserted that "we will not give the Arctic to anyone."[16] Although his remarks do not correspond with the legal position of the Russian state, whose claims strictly respect international laws, they have never been rejected by Putin, who was happy with the provocative character of Chilingarov, whose declarations were essentially addressed to a domestic audience.

The will to turn the Arctic into a component of the state-centric patriotic narrative became even more apparent in 2009 in light of the decision to revive the Russian Geographical Society, itself born in 1845 as part of the imperial drive for geographical expansion and exploration of the country's natural resources, and to turn it into one of the Arctic science flagships.[17] Then Minister of Emergency Situations Sergey Shoigu was appointed its president, while Putin assigned himself the post of chairman of the Council of Trustees. Putin has not concealed his desire to have the activities of the Geographical Society focus on his main state-sponsored projects: "The Society can offer practical support to our plans to develop East Siberia and the Far East, Yamal and the north of the Krasnoyarsk region, to participate actively in further research projects in the Arctic and Antarctica, as well as environmental support of the Olympic Games in Sochi."[18]

Arctic Exceptionalism Challenged by the War in Ukraine

Even the first conflict with Ukraine in 2014, with the annexation of Crimea and a Moscow-backed insurgency in Donbas, did not destabilize Arctic cooperation projects, as all countries involved wanted to maintain this exceptionalism.[19] Of course, the region did not escape from recurrent rhetorical tensions and the proliferation of military exercises, which were mutually interpreted as provocations, casting a shadow over local partnerships. Scandinavian countries, in particular, complained about increasingly frequent military incursions into their airspace and maritime areas by Russian bombers and ships.[20] In 2018, Sergei Shoigu, by this time the Russian defense minister, declared that competition in the Arctic could lead to potential conflict.[21] But the "friend-enemy" logic still allowed regional and bilateral structures to function.[22] In 2021, when Russia assumed the chairmanship of the Arctic Council for a two-year rotating term, it announced an active presidency focused on issues of sustainable development and collective governance.[23]

However, Arctic exceptionalism could not withstand the seismic event of Russia's full-scale invasion of Ukraine on February 24, 2022. Moscow wanted to continue separating the Arctic from other international tensions, but Western countries refused to let it. On March 3, 2022, two weeks after the invasion began, all Western member states suspended their participation in the Arctic Council. Some voices even suggested the possibility of a complete dissolution of the institution. By June, however, the states decided to continue their work on climate change and sustainable development in committees where the vote of all members is not required, so as not to involve Moscow.[24]

At the end of 2023, the eight states agreed on a mechanism that allows the working groups—the core of the Council—to be relaunched, even though it is difficult to imagine a return to normalcy where Russia could vote again as a full member in the near term. In February 2024, the Norwegian presidency of the Arctic Council announced the gradual resumption of scientific activities. The various treaties signed under the auspices of the Council, particularly the SAR Agreement and the Marine Oil Pollution Preparedness and Response Agreement (MOSPA), remain in force, even though, in practice, cooperation is frozen, endangering the overall security of the Arctic.[25]

Russia's bilateral relations with each of the Scandinavian countries have also greatly suffered: cross-border exchanges that had transformed regions like Tromsø and Kirkenes in Norway, or Nikel and Zapoliarnyi in Russia, were gradually halted, penalizing local populations. Tensions have multiplied around the use of migration as a political pressure tool. Finland and Norway have accused Russia of sending hundreds of asylum-seekers across Arctic borders, and in 2023 and 2024 they decided to build a fence at their borders to deter any illegal crossing.[26] Furthermore, cooperation between Russia and the EU, Norway, and Iceland, known as the Northern Dimension, has been halted, while two other regional cooperation institutions, the Barents Euro-Arctic Council and the Council of the Baltic Sea States, have suspended Russia's participation, before Moscow decided to cease membership entirely.

However, the main international treaties remain in force, such as the Svalbard Treaty, which regulates international presence in the archipelago under Norwegian sovereignty, including the small Russian mining town of Barentsburg, and the joint management of fishing zones between Norway and Russia, which was renewed at the end of 2023. Oslo indeed seeks to maintain a minimal level of institutional cooperation and does not wish to sever all ties with Russia. Most treaties in force in the Arctic are multilateral treaties linked to UN institutions or have a global dimension, which protects them from the Ukraine-centered geopolitical crisis: the UN Convention on Climate Change, the Polar Code of the International Maritime Organization, the 1973 Agreement on the Conservation of Polar Bears, the Agreement to Prevent Unregulated High Seas Fisheries in the Central Arctic Ocean, also signed by Asian countries such as Japan, South Korea, and China, are all still in effect.

Since then, even though Russia continues to portray the Arctic as a region of peace, stability, and international cooperation,[27] its tone has also changed. None of the Arctic regional structures were mentioned in the Foreign Policy Concept published in 2023.[28] The same year, amendments to the official document, "Russian Arctic Strategy until 2035," announced that Moscow would henceforth develop its Arctic relations on a bilateral basis and with sole consideration of its national interests.[29] In February 2024, the Russian Ministry of Foreign Affairs even invoked the possibility of leaving the Arctic Council in the event it can no longer perform its duties[30]—a threat that would negatively impact Russia itself, too, as the Council remains a place for Moscow to defend its pro-multipolarity position, especially as the Global South becomes gradually more involved in Arctic affairs. The circumpolar region has thus now been divided into two Arctics isolated from each other: a NATO Arctic with seven members and a Russian Arctic cut off from its circumpolar neighbors but open to the Global South (see further).

National Sovereignty and the Militarization of the Arctic

The sector most affected by the ripple effects of the war in Ukraine has undoubtedly been the military domain. However, what is defined as militarization in the Arctic needs to be put into perspective. There are no conflicts between Arctic states over their respective territories, and the prospective issues of continental shelves are negotiated diplomatically through UNCLOS, as seen in Chapter 1. The modernization of nuclear equipment is part of the deterrence mechanisms inherited from the Cold War years and the strategic balance between Russia and the United States. At least until the 2022 crisis, Russian doctrines and Vladimir Putin's speeches confirmed that strictly military risks were not considered by Moscow as a real danger for the Arctic region compared to civil security risks.

The gradual remilitarization of the Arctic by Russia should therefore be understood mostly as the reaffirmation of national sovereignty over a region that was neglected by the state in the 1990s and 2000s (see Map 2.1). First among Moscow's Arctic priorities is the Kola Peninsula, home to most of Russia's submarine ballistic nuclear missile launching ships ("ship submersible ballistic nuclear," or SSBNs) capable of nuclear response, and therefore central to the country's nuclear deterrence and second-strike capability. Second is the security of the Northern Sea Route and its adjacent continental territories. Third is the staging ground for power projection and multilayered sea denial, especially into the North Atlantic Ocean, thanks to access through the Barents Sea. The latter is Russia's unavoidable route to the international waters of the Atlantic Ocean via the GIUK gap (the maritime choke point between Greenland, Ireland, and the UK), which gives the Russian Navy access to the Atlantic and has seen a renewal of submarine activity.

While the Arctic was a highly militarized zone during the Cold War decades, Russia's strategic infrastructure collapsed in the 1990s. The Northern Fleet was then in poor condition, and the nuclear infrastructure of the Kola Peninsula was maintained only minimally to ensure nuclear deterrence and prevent environmental disasters. Images of abandoned, rusting Russian submarine carcasses have been circulating in the media and on the internet, and the risks of pollution, including radioactive pollution, from these open-air dumps of military equipment have been an issue of serious concern, especially for Russia's neighbors such as Norway.[31] Vladimir Putin gradually revitalized the military sector and refinanced a humiliated and impoverished army during his first two terms (2000–8), but the Arctic was not significantly affected by this remilitarization until the 2010s.

Russian military presence in the Arctic slowly resumed in 2007, with Moscow patrolling NATO borders with strategic bombers. However, compared to the Baltic or Black Seas, detection of Russian bombers in the region by NATO radars remained well below Cold War standards. It was only in 2014 that the region was explicitly mentioned in the Russian military doctrine. Tensions with the West were escalating. The five NATO member states of the Arctic Council ceased their military exchanges with Russia, even halting the circumpolar SAR mechanism for some time before restarting it. Since the past decade, in an implicit mirror game, both Russia and NATO countries have increased their military activities in the Arctic, conducting sophisticated military exercises there.[32] To give but one example, in 2021 Russia sent three nuclear submarines to the Franz Joseph Archipelago, Russia's northernmost territory, to showcase its Arctic sovereignty. On the other side, in 2024, fourteen NATO countries and partner countries organized Nordic Response, a massive military exercise gathering of over 20,000 soldiers, to exhibit their Arctic military preparedness.[33]

To institutionalize the upgrading of the Arctic region into Russia's strategic projection, the government decided in 2020 to separate four northern regions—Murmansk, Arkhangelsk, the Komi Republic, and the Yamalo-Nenets Autonomous District—from the Western Military District of Russia and integrate them into a new unified strategic command of the Northern Fleet. It did become an autonomous military district in 2021, yet this decision was canceled in 2024 in the context of the reorganization of the Russian Army on a war footing.[34] Nonetheless, the Arctic has remained a central element of Moscow's strategic portfolio, visible especially in the steady modernization of the Northern Fleet.

Based in Severomorsk, near Murmansk, the Northern Fleet embodies the Russian Navy's mission to be present in the "global ocean," including up to the Red Sea and the Indian Ocean (and indeed in 2023, the frigate *Admiral Gorchkov* sailed from South Africa to Saudi Arabia). The Northern Fleet therefore positions itself as a central symbol of Russia's great power status on display for the Global South to see, with this paradox of being an Arctic fleet often navigating outside its designated environment.[35] For a long time, the Navy remained a big loser in Russia's military budget programming, with insufficient funds provided to renew ships, whose lifespan has often been overstretched.[36] In 2019, three incidents revealed the lack of funding for much Arctic infrastructure and the existence of security risks that were often underestimated: the explosion on board an AS-31 nuclear submarine, the explosion of a nuclear-powered Burevestnik missile, followed by an explosion on one of the Rosatom sites in Nionoksa, near Arkhangelsk.[37]

But over time, the renewal of the Northern Fleet's infrastructure bore fruit, with the arrival of new submarines, frigates, and long-range missiles such as Kalibr and Tsirkon. The Northern Fleet has now become one of Russia's best multidimensional naval assets: it combines naval forces (submarines, large surface ships, landing ships, corvettes and patrol vessels, minesweepers), air forces of different capacities (including anti-submarine aircraft and anti-submarine helicopters), land forces, as well as a developed toolkit for what is called "hybrid warfare" (to include underwater reconnaissance, electronic warfare, potential pipeline and cable sabotage, etc.). Since 2022, it has also received new technologies such as the nuclear-powered cruise missile Burevestnik, which can carry nuclear warheads; new Borey class submarines; and nuclear unmanned underwater vehicle (UUV) called Poseidon.[38]

The Northern Fleet has been redeployed to the Black Sea for the Ukrainian War, and it is said to have suffered extensive losses in the first months of the 2022 invasion of Ukraine.[39] War has shown the fragility of the Russian fleet in the Black Sea, often attacked by Ukrainian drones up to the point of several ships sinking. Yet the Northern Fleet has maintained its normal level of activity in the Far North and especially the Barents Sea, a sign that for the Kremlin, the Arctic remains a strategic region that cannot afford to be left unattended.

To accompany this renewed activism in the Arctic, the Russian authorities have also gradually reopened eight air bases (most of which had been out of use since the fall of the USSR), in addition to opening six new military bases along the Northern Sea Route, each hosting between 150 and 600 personnel.[40] Among these, three are fully autonomous and equipped with long-, medium-, and short-range missiles. The main base is the Nagurskoye Air Base, Russia's northernmost military installation, located on the Franz Josef Land Archipelago. This spectacular "Arctic Trefoil" (*Trilistnik*), emblematic for its modernist look in the colors of the Russian flag, consists of brand-new buildings and can accommodate MiG-31 and SU-24 aircraft, bringing the American coasts within range of Russian bombers. Temp Air Base, on Kotelny Island, also operational since 2015, can host large aircraft such as the Ilyushin Il-76, as can the Rogachevo Air Base on the Novaya Zemlya Archipelago. The other bases at Cape Schmidt, Wrangel Island, and Sredniy are more modest in size.

This infrastructure has been accompanied by the formation of new Arctic brigades. While trained primarily for civil security missions, these Arctic elite troops are equipped with state-of-the-art gear and capable of handling extreme situations in the polar environment.[41] Globally, the military and security services are responsible for patrolling the Arctic national territory, and given the region's exceptional climatic conditions and isolation, the technologies, equipment, and specialists there are often dual use, both military and civilian.[42] Indeed, the military is the only force capable of performing civil security missions along the Northern Sea Route, having the logistical capacity to organize SAR operations in case of natural or industrial disasters (as well as in case of a circumpolar flight accident), combating illegal overfishing, countering potential acts of sabotage on critical infrastructure, and assisting scientific research, particularly in meteorology and oceanography.

Given the level of tension between the West and Russia following the invasion of Ukraine, the accelerated militarization of the Arctic now seems difficult to avoid. In June 2022, NATO Secretary General Jens Stoltenberg declared that the Arctic was of strategic importance to the alliance. Russian Minister of Defense Sergei Shoigu stated in December that new military bases would be established: "Given NATO's desire to build up military potential near the Russian borders, as well as to expand the North Atlantic Alliance at the expense of

Finland and Sweden, retaliatory measures are required to create an appropriate grouping of troops in Northwest Russia."[43] In its March 2023 Foreign Policy Concept, the Russian government argues the Arctic is a region of "special interest" and that its mission is to "neutralize the militarization movement of the Arctic by unfriendly states."[44]

While the risk of a direct military conflict in the Arctic seems low, a global escalation linked to the war in Ukraine is no longer purely theoretical. The risk of a misinterpreted provocation or an accident raises concerns. Western countries have been for instance denouncing increased GPS jamming by Russia, which puts civilian aviation at risk, along with attacks on electrical stations and undersea cables.[45] The sinking of the *Ursa Major*, a Russian cargo ship, in the Mediterranean Sea on December 24, 2024, is symptomatic of the exacerbation of maritime tensions in the wake of the Ukraine War. The ship, belonging to the Russian Ministry of Defense and under US sanctions, was carrying crucial elements for the Arctic program: metal components from Saint Petersburg and powerful cranes for the Zvezda shipyard near Vladivostok, where the new icebreakers for the Northern Sea Route are being built.

With the entry of Finland and Sweden into NATO, the NATO-Russia interface in the region has been completely transformed.[46] The division between the Arctic as a space of cooperation and the Baltic as a space of competition is largely disappearing, and the two regions are now directly connected via Finland and Sweden joining NATO. Coordination among NATO countries in the Arctic, previously dominated by the United States and Norway, will be strengthened with greater interoperability among allied armies now at Russia's doorstep. The two new members, Finland and Sweden, will open their Arctic military bases to American forces, which have also reactivated a base for 12,000 troops, the 11th Airborne Division, in Alaska.[47] For a long time, the United States was reluctant to visibly engage in the Arctic, but it opened a diplomatic post in Tromsø, in the north of Norway, at the end of 2023. The US Department of Defense, following several military exercises in the Arctic, is set to publish a new Arctic Strategy in 2025.[48] The strategic position of Russia has thus become more complicated with NATO's new expansion along its borders as well as with the Trump administration's very vocal comments on "buying" Greenland from Denmark.

Nuclear Modernization and the Icebreaker Fleet

The nuclear aspect of this militarization of the northern lands has been present since the beginning of the Cold War. From 1954 until 1990 (the date of the moratorium on nuclear tests in the Arctic), a nuclear test site operated on the southern island of the Novaya Zemlya Archipelago, with 132 tests of various types conducted there. In October 2023, Vladimir Putin suspended Russia's participation in the Comprehensive Test Ban Treaty, signed in 1996 but never fully ratified by the United States. Officially, Moscow says it reserves the right to resume testing if the United States does the same. In September 2024, the head of the Novaya Zemlya nuclear test range declared his site ready for the resumption of open-air tests, should the president give the order.[49] This is part of the new strategic state of affairs in the Arctic, wherein Cold War-inspired nuclear deterrence and signaling has become the new normal.

Civilian nuclear infrastructure has also proliferated in the Arctic. Russia has developed many radioisotope thermoelectric generators (RITEGs), small nuclear power generators deployed by the hundreds along the Northern Sea Route to power lighthouses, beacons, military relays, and weather stations. Rosatom commissioned the first floating mini nuclear power plant barge, the *Akademik Lomonosov*, in Pevek in 2019, in order to supply electricity to parts of Chukotka. In 2024, it inaugurated a new mini nuclear power plant in Sakha-Yakutia near important gold and other mineral mines.[50]

At the heart of the dual military-civilian nature of Russian Arctic nuclear infrastructure is the fleet of icebreakers managed by the state nuclear corporation Rosatom and its maritime subsidiary, Rosatomflot. Russia leads the world in icebreaker production, with eighty-five in operation and twenty-one under construction as of January 1, 2024, far ahead of the twenty-three in use by Canada, sixteen for the United

States, and five for China.[51] Russia is also the only country capable of producing nuclear icebreakers, with seven currently in operation along with one unique nuclear-powered container ship.

However, despite this status as a global leader, Russian production does not meet Moscow's ambitions for the Northern Sea Route. A new nuclear icebreaker, the *Chukotka*, was put into service in 2024 by the shipyards of the state-owned United Shipbuilding Corporation in Saint Petersburg, with three more—the *Yakutia*, the *Leningrad*, and the *Stalingrad*—currently in production.[52] Even more powerful icebreakers, nicknamed "Leaders," have been in production at the Zvezda shipyard (in the Far East Federal District), which belongs to Rosneft, the private oil company headed by Igor Sechin, a close associate of Putin. Still, sanctions are weighing on Russian access to Western technology, and instead of two Leader nuclear-powered icebreakers, only one is planned to be built as of 2025.

As always, construction has faced multiple challenges, resulting in several years of delays. Cost overruns and technological complications are due in part to Zvezda's lack of experience in building icebreakers, insufficient production capacity, and a shortage of specialized labor. In 2022, Rosneft managed to take control of the Russian company Iceberg Central Design Bureau, one of the world's leaders in icebreaker design, to advance its project. With the military invasion of Ukraine, Russia also lost access to the Ukrainian plant Energomashspetsstal in Kramatorsk, which it bombed while it was producing parts necessary for its icebreakers. Now, Russia must find ways to replace and/or produce the 15 to 20 percent of Western- or South Korean-sourced spare parts that are under sanction.[53]

These difficulties are representative of the systemic problems of the Russian economy, which must balance contradictory demands. In the case of icebreakers, Moscow wanted to relocate the shipyards, which were too close to the country's western borders, to the Far East for strategic reasons, and to help Rosneft establish its oil and gas bases in Asia (with major sites being Sakhalin 1, 2, and 3). This move, however, came at the expense of leveraging the expertise acquired by the traditional Russian shipyards in Saint Petersburg and in Severodvinsk, near Arkhangelsk.

Western sanctions have compounded these systemic problems, forcing Russian companies to seek new partners or acquire the missing technological know-how themselves. The absence of polar-class bulk carriers needed for coal transport, due to sanctions, also pushed Moscow to authorize non-reinforced hull tankers to operate along the Northern Sea Route, posing significant environmental risks.[54] The Russian government has announced a new wave of investments in Arctic ports, but very few international shipping companies are willing to commit, primarily due to the exorbitant cost of insurance, as international insurers are no longer willing to cover Russian territory.

China: Navigating Alliance and Competition in the Arctic

The western facade of the Arctic dominates Russia's strategic and economic concerns: the majority of both commercial and military traffic occurs between the Barents Sea, with the latter constituting the military heart of the Russian Arctic, the White Sea, and the Yamal Peninsula. However, Moscow has gradually turned toward Asia in search of new partners, hoping to revitalize the eastern part of its Arctic, which is much less developed (particularly from Tiksi, the small port in northern Sakha-Yakutia, to Chukotka). As in the rest of Russian foreign policy, the lion's share of investments goes to China.

China has steadily engaged in Arctic affairs, initially through scientific activities. It claims the existence of a "third pole," the Tibetan Plateau, whose glaciers and permafrost are quite similar to the Arctic and Antarctica, to project itself as a "near-Arctic" power.[55] Beijing has also sought to export its "infrastructure diplomacy" and aggressively positioned itself throughout the Arctic region, not only in Russia but also in Iceland and Greenland, with a discourse on the "Polar Silk Roads."[56] It has become Alaska's largest export market; bought shares in mines in Canada and Greenland; engaged in space observation activities in Iceland, Sweden, Finland, and the Svalbard Archipelago; and voiced its own interests at the International Maritime Organization. The terminology of being a "near-Arctic" power, used in China's first Arctic White Paper of 2018, was contested by many

Russia's Arctic

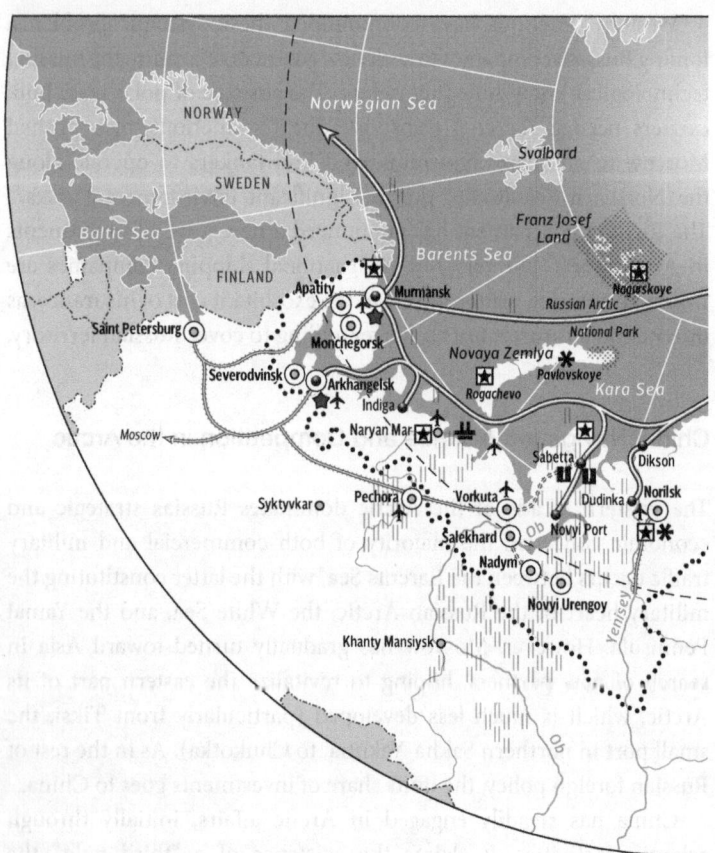

Map 2.1 Economic and strategic development in the Russian Arctic.

Major Diplomatic and Security Space

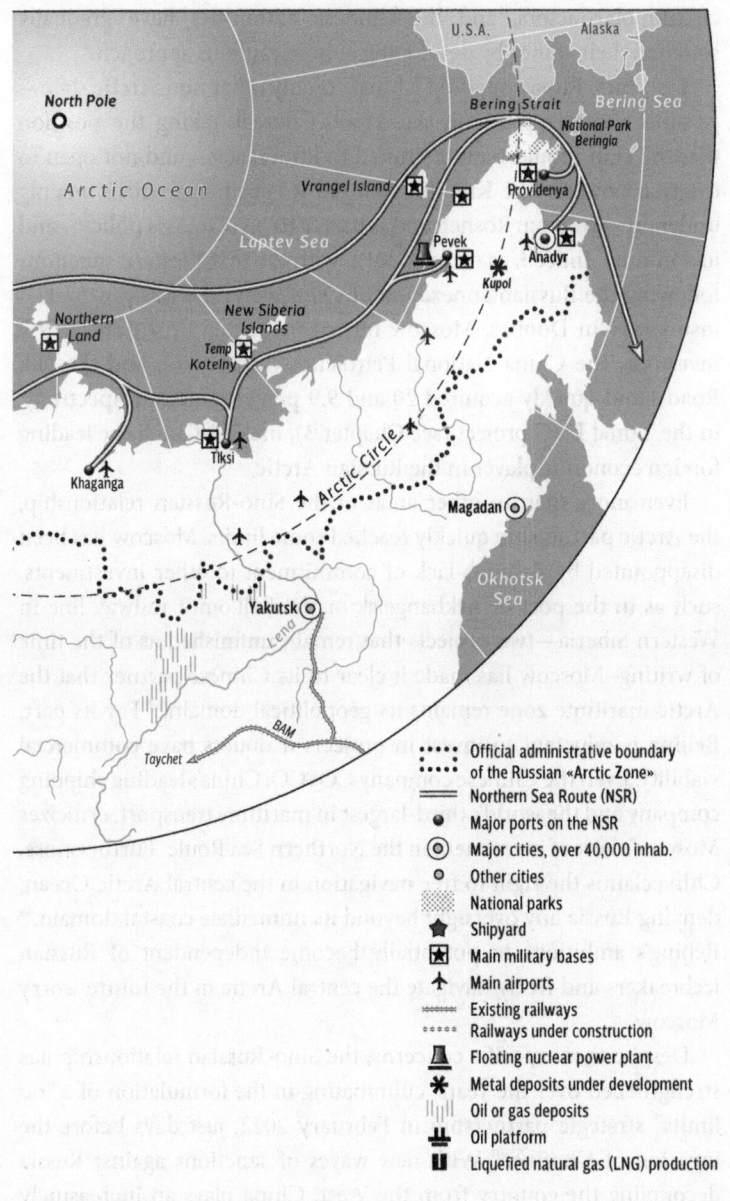

49

circumpolar actors, and the Chinese authorities have gradually softened their stance by developing a more cautious approach.[57]

For years, Russia opposed China—or any other non-Arctic state—gaining observer status in the Arctic Council, taking the position that the club should remain limited to littoral states and not open to external powers. The Kremlin changed its position in 2013, mainly under pressure from Rosneft and as part of its pivot to Asia policy—and just in time. Indeed, as early as 2014, with the first Western sanctions following the Russian annexation of Crimea and the Russian-backed insurgency in Donbas, Moscow turned to Beijing in search of new investors. The China National Petroleum Corporation and the Silk Road Fund quickly acquired 20 and 9.9 percent stakes, respectively, in the Yamal LNG project (see Chapter 3), making China the leading foreign economic player in the Russian Arctic.

Even more than in other areas of the Sino-Russian relationship, the Arctic partnership quickly reached some limits. Moscow has been disappointed by Beijing's lack of commitment to other investments, such as in the port of Arkhangelsk or the Belkomur railway line in Western Siberia—two projects that remain unfinished as of the time of writing. Moscow has made it clear to its Chinese partner that the Arctic maritime zone remains its geopolitical domain.[58] For its part, Beijing is reluctant to invest in projects it doubts have commercial viability. Even the Chinese company COSCO, China's leading shipping company and the world's third-largest in maritime transport, criticizes Moscow's lack of investment in the Northern Sea Route. Furthermore, China claims the right to free navigation in the central Arctic Ocean, denying Russia any oversight beyond its immediate coastal domain.[59] Beijing's ambitions to potentially become independent of Russian icebreakers and freely navigate the central Arctic in the future worry Moscow.

Despite these specific concerns, the Sino-Russian relationship has strengthened over the years, culminating in the formulation of a "no limits" strategic partnership in February 2022, just days before the invasion of Ukraine.[60] With new waves of sanctions against Russia decoupling the country from the West, China plays an increasingly significant role in the Far North. Even though Chinese companies have abandoned their shares in Arctic LNG 2 due to fears of US secondary sanctions, they remain very much present in the region. For example, they have delivered the necessary equipment for the construction of LNG turbines, agreed to invest in the Pizhenskoye titanium mine in the Komi Republic, and supported Gazprom in constructing petrochemical plants for the Yamal-Europe Pipeline. The China Communications Construction Company (CCCC) is set to engage in mineral exploitation with Russian Titanium Resources and invest in the deepwater port of Indiga in the Nenets Autonomous District. Additionally, Nornickel is negotiating the transfer of one of its copper smelting plants to China to avoid sanctions and maintain access to international markets.[61]

It appears that in less visible areas than the major gas deposits, Chinese presence has even increased: the American firm Strider, which aggregates open data, has identified 234 Chinese companies authorized to work in the Russian Arctic between January 2022 and June 2023, an increase of 87 percent compared to 2021.[62] Chinese inroads have also been noticeable through the multiplication of direct contacts between, for instance, the governor of the Murmansk region and the Chinese company Shandong Port Group (as well as for tourism-related purposes), and the governor of the Arkhangelsk oblast and the authorities of the Chinese port city of Dalian to discuss Chinese investments in Russian cities' infrastructure and logistics.[63]

China indeed continues to be very interested in the Northern Sea Route, which guarantees a geopolitically secure (though logistically challenging) passage as tensions with the United States in the South China Sea escalate and Beijing worries about a possible maritime blockade. COSCO has been exploring the use of the Northern Sea Route, and in May 2024 both countries decided to create a joint commission for the development of the route.[64] Following a trial voyage in 2022, eleven ships transported Russian oil to China through it in 2023, with a combined forty-four from Russia to China and from China to Russia (mostly minerals) in 2024.[65] The Chinese People's Liberation Army Navy does not yet have significant Arctic capabilities, and its two polar ships, *Xuelong 1* and *2* (meaning *Snow Dragon*), are primarily used for scientific operations. However, Chinese shipyards have gradually developed a wide range of ice-class vessels in various domains, including icebreakers, mineral cargo ships, and tankers. And at the 2024 Saint Petersburg Economic Forum, a Chinese shipping company even signed a major contract to launch a container shipping line via the Northern Sea Route.[66] For Chinese companies, the development of this alternative route is also a favorable factor for the growth of ports in northern China, which are less well positioned in relation to the main shipping lanes running to South Asia.

But this Chinese commitment is by no means absolute. Indeed, even though China continues to support Russia in the face of Western sanctions, it has refused to write a blank check for investment in Russia's Arctic regions.[67] The Chinese heavyweight is itself grappling with a post-Covid economic crisis, and it has access to just about all the Russian natural gas it needs already. Moreover, Chinese companies, especially large ones with a presence in global markets, are keen to avoid the risk of US secondary sanctions.[68] In the medium to long term, however, it is conceivable that Chinese presence in the Far East will gradually move northward—a trajectory that has long interested the authorities of Sakha-Yakutia, who would like to see Chinese investments in the republic's infrastructure to connect it to the markets of the Far East.[69]

Beijing has also become more involved in the satellite domain, providing Russian meteorological services with real-time information on ice conditions along the Northern Sea Route.[70] Militarily, Russia has never been interested in involving its Chinese ally in this key area of national sovereignty, and bilateral Russo-Chinese military exercises have never included an Arctic component. However, this is changing as well, with the two countries conducting joint naval exercises in the international waters of the Bering Sea and sailing together near Alaska's Aleutian Islands in 2022, 2023, and again in 2024.[71] This strategic cooperation could be strengthened by an agreement signed in 2023 between the FSB and the Chinese Coast Guard to enhance the enforcement of maritime laws—an agreement that signals a shift in Russia's policy of avoiding any involvement of a foreign country in its sovereign domain.[72]

Enter the Global South: Moscow in Search of New Partners

While China is obviously Russia's largest economic and strategic partner, it is not the only Asian country looking toward the Arctic. As Lukas B. Wahden explains, Russia serves as the gateway to the Arctic for so-called Global South countries that are seeking new avenues for international influence and new partnerships.[73] For its part, Moscow is looking for new allies to diversify its investors and avoid being trapped in dependence on China.

Since 2022, India has become one of the main clients for Russian oil and coal, overtaking China as Russia's top oil buyer. In December 2024, Rosneft signed its biggest oil deal ever, a ten-year agreement to supply 500,000 barrels of crude oil per day to Indian private refiner Reliance.[74] In July 2024, Vladimir Putin and Indian Prime Minister Narendra Modi signed a trade and cooperation agreement for the Far East and the Arctic up to 2029. This was followed a few months later by the meeting of the Indian-Russian group on cooperation in the Northern Sea Route to discuss the growth of Indian-Russian cargo transit, joint projects in the Arctic shipbuilding, and possible training of Indian sailors for polar navigation.[75] Both countries have also mentioned the International North-South Transport Corridor (a multimodal freight network of ship, rail, and road between India, Iran, Azerbaijan, Turkmenistan, Kazakhstan, and Russia) and the Vladivostok-Chennai maritime route. Several Indian companies are now investing in Arctic cities, such as Saafarm building a pharmaceutical production facility in the Murmansk region.[76]

This Russo-Indian cooperation goes beyond energy ties. India published its own Arctic strategy in 2022 and sees itself (similarly to China) as a "tri-polar" nation, present in the Arctic, Antarctica, and the "third pole" that is the Himalayas. As a country seriously concerned about climate change and especially coastal erosion, India is also interested in acquiring scientific knowledge on climate change. The Russian Arctic and Antarctic Institute and the Indian National Center for Polar and Ocean Research already cooperate in Antarctica and envision developing a research base in the Arctic, "Northern Pole"

(Severnyi polius).⁷⁷ Moscow and New Delhi also have a long tradition of space cooperation between Roskosmos and the India Space Research Organisation (ISRO), with plans to put up satellite receiving stations in the Arctic and develop digital connectivity in the Arctic.⁷⁸

Russia has also approached Turkey, Vietnam, and the United Arab Emirates for technological and naval cooperation in the Arctic in hopes of circumventing Western sanctions. Dubai's DP World, one of the globe's largest logistics companies, inked a deal to establish a joint venture to develop container shipping along the Northern Sea Route.⁷⁹ Moscow is considering a direct railway line between Murmansk and Bandar Abbas, in Iran—a project already conceived in the 2000s and still not made a reality—so that Russian cargo could gain direct access to the Indian Ocean. At the 2024 BRICS summit in Kazan, Rosatom signed new deals to sell its floating electricity stations, built for Arctic conditions but adaptable to warmer climates, to its Global South partners.⁸⁰ Last but not least, Russia's "shadow fleet" of tankers, which bypass the Western sanctions and oil price-cap framework, has operations in Africa, Southeast Asia, and North Korea.⁸¹

Far from being purely economic in nature, the rise in visibility of the Global South in the Far North is also diplomatic and cultural. Russia, for example, is considering transforming the two small Soviet mining towns of Barentsburg (now down to only 400 inhabitants) and Pyramiden (which has been entirely abandoned) on the Svalbard Archipelago into showcases for its scientific cooperation—and tourism—with the BRICS. Against the opinion of the Norwegian government, Moscow would like to build a new BRICS scientific facility linking natural and social sciences and competing with the Svalbard Global Seed Vault, the world's largest backup facility for seeds and a major Arctic scientific platform.⁸² Diplomatically, Moscow seems to be adopting an open-door policy toward Global South countries in the Arctic, with the aim of weakening the unity of Western states in general, and all NATO members in particular, especially those within the Arctic Council. Consequently, Russia could become, as in other areas, a key player in the international shift toward the Global South, with implications reaching into the Arctic.

CHAPTER 3
SUCCESSES AND CHALLENGES OF ARCTIC ECONOMIC TRANSFORMATIONS

Since the end of the Cold War, the Arctic has become one of the key assets for Russian economic growth. Including the sub-Arctic Zone, it produces about 80 percent of the nation's gas; 60 percent of its copper; 17 percent of its oil; more than 90 percent of its nickel, cobalt, and main rare earth metals; 96 percent of its platinoids; and 100 percent of its barite and diamonds.[1] While military issues contribute to reinforcing the importance of this zone among the government's priorities, economic factors remain all the more crucial since, with the depletion of hydrocarbon deposits in the Middle Ob, the share of new deposits in the Far North, particularly Yamal, is growing.[2] Moreover, the rupture with Russia's main partners following Western decisions to impose sanctions and boycott Russian hydrocarbons has shattered the system established since the 1960s for supplying European customers via pipelines. This represents an unprecedented challenge for today's Russian economy, one of whose key elements could be the accelerated development of this unique maritime facade, with the new role assigned to liquefied natural gas and its export to Asian clients via the Northern Sea Route.

The Laborious Development of a Region with Multiple Constraints

During the first half of the twentieth century, the resources exploited in the Arctic region were limited to minerals (polymetals on the Kola

Peninsula and Norilsk; silver, gold, and tin in Yakutia and Kolyma; diamonds in Yakutia), timber and coal in the Vorkuta region, and some fisheries along the Northern Sea Route. Alexander Piliasov, one of Russia's leading experts on the Far North, distinguishes four phases in this development. The first phase, from the 1930s to the 1950s, was marked by the Gulag camps and their servile labor force equipped with shovels and pickaxes and subjected to dramatic living conditions. The second phase, from Stalin's death in 1953 until the end of the Soviet era, was characterized by sectoral administrations (different administrative institutions were supervising different sectors) and large planned industrial complexes that managed the exploitation of deposits, the construction of accompanying towns and factories, and their entire social environment. It was within this framework that hydrocarbon deposits in the Ob Basin began to be exploited from the late 1960s onward.[3]

The third period, the 1990s, was one of great upheaval. Under Boris Yeltsin, the state privatized most of the relevant industrial and mining sectors (except for gas), to the advantage of oligarchic-run corporations such as Yukos, Lukoil, Nornickel, and Rosneft. While hydrocarbon-producing regions became oriented toward exports, they managed to maintain a relatively stable level of production during this period, while most other regions went into crisis mode. According to Piliasov, this was a true deindustrialization, resulting in the collapse of most non-extractive modes of production and even the abandonment of entire sectors such as food industry, construction materials, or various industries serving the local population. As the costs of energy, transportation, and services soared, the state no longer had the means to finance these sectors or the bonuses for the population, which preferred to return to southern Siberia or the European part of the country (see Chapter 4).

The final phase began with the rise of Vladimir Putin to power in 2000, characterized by a clear economic recovery on multiple fronts. Even when it did not proceed with outright renationalizations, the state regained control of major mining sectors by pressurizing the oligarchs regarding their financing choices and commercial orientation, particularly the export share burdened with specific taxes.[4] The government reinstated an investment policy on this strategic front. While prioritizing large transportation infrastructure projects such as the Northern Sea Route, it multiplied public-private partnerships and promoted dual-use projects like ports in the Murmansk region or Novaya Zemlya, serving both military and economic functions.[5]

To overcome Russia's technological lags and attract capital, Putin first amplified the policy initiated by Gorbachev and Yeltsin, favoring partnerships with major Western firms. This led to the beginning of a significant dual reversal in the geographical orientation of Arctic development. While most projects since the 1930s had been conceived in the south to be implemented in the north, including regarding the direction of sending production, increasingly operations were being carried out from the north itself, relying on the Northern Sea Route. At the same time, particularly since the full-scale invasion of Ukraine in February 2022, efforts have been made to favor direct exports to Asian countries from various developing ports along this maritime route.

A Strategic "Granary" of Minerals

The exploitation of the Arctic's mineral wealth began in the 1930s through Gulag projects. Over the years, geologists discovered numerous new deposits in various domains, as detailed in a remarkable report published in 1983 by three American authors.[6] As Soviet planners of the time proclaimed, the USSR possessed deposits on its territory covering nearly all the elements of Mendeleev's periodic table. Initially, this involved coal and metallic or precious metal products, as hydrocarbons were not discovered until the late 1950s.

However, the constraints of exploitation presented numerous obstacles to development. First, there was what is known as the "cold tax": the need to use machinery and materials capable of withstanding extreme temperature fluctuations, whether it be the steel of machines and pipes or the concrete used in construction. Such specially adapted equipment is necessarily more expensive than that used in the European part of the country.[7] Furthermore, most of the natural resource deposits in the Arctic Zone are located in what the Russians refer to as "roadless areas" (*bezdorozhnye*), meaning there are no practical means of access, except sometimes via river routes usable during the freezing period (where ice roads, known as *zimniki*, are constructed) or during the thaw. But the development of such large metal ore deposits requires heavy infrastructure, often with on-site refining units. As a result, only a few truly strategic sites, such as Norilsk (which has nickel, platinum, copper, cobalt, and palladium), have been developed, while most others have been put on hold for decades, even if their reserves are significant.

One region in the Arctic Zone, the Kola Peninsula, partially escapes from these constraints due to its geographical position (its northern coast is warmed by the Gulf Stream's final offshoot, the Norwegian Current) and its early development. Rich deposits of apatite, iron, and nonferrous metals were discovered there early on. It also has a network of industrial towns connected by roads and railways. After the fall of the Soviet system, the region faced the problem of depleting some deposits and the obsolescence of many mines or enrichment plants.

Since the 2000s, several programs have been launched to modernize the mining sector. The most important of these are the Kovdor (iron ore and apatite) and Kirovsk combine; Apatity, founded in 1929 in the Khibiny Mountains, is one of Russia's leading suppliers of nepheline and apatite, used in the production of fertilizers and phosphate-based technical compounds. Its production fell sharply from 20 million tons of apatite concentrate in 1988 to less than 5 million tons in the early 2000s. The use of poorer deposits or rejected materials from previous operations, and the installation of new, more environmentally friendly equipment, helped grow the production of apatite and nepheline to 25 million tons per year, and revive the town of Kirovsk and other single-industry towns in the region.[8] Already favored by its strategic position and access to sea lanes and rail links, the Murmansk region is the focus of numerous investments designed to consolidate its economic and military importance. A railway bridge project will link the two shores of the Kola Estuary and favor the development of new coal and hydrocarbon terminals at the port of Lavna on the left bank.

Arctic Economic Transformations

Some major sites received original solutions. In 1959, the Cherepovets steel plant north of Moscow was inaugurated with dual Arctic supplies. Iron came from the Kola Peninsula, and coking coal from Pechora via specific railway routes. The scheme used in Norilsk since the 1940s has been more complex. The metals enriched on-site are transported by a special railway to the port of Dudinka on the Yenisei, where they are stored until the navigable season, then shipped via the Northern Sea Route to the western part of the country. The city and its factories faced a difficult period after the fall of the USSR. With the decline in state orders, debts accumulated, and the combine could no longer maintain the operation of this city, which had been a symbol of the Soviet will to defy natural constraints.

Norilsk is now known as one of the most polluted cities in the world. Breaking with previous voluntarism, the city council proposed reducing the population by sending retirees and nonessential families back. The city lost 30 percent of its inhabitants between 1985 and 2001 (see Chapter 4). In 1993, the combine was privatized and purchased two years later by Vladimir Potanin's ONEXIM Bank. In 2016, he closed the Nikelevyi plant (created in 1942), the oldest and most polluting of the enrichment units. Some refining was transferred to Monchegorsk, a modern plant on the Kola Peninsula where semi-processed metals are shipped via the Northern Sea Route using a special fleet acquired by Nornickel. Gradually, the city's population stabilized, and the various units were modernized to reduce pollution, but the environment remained degraded for miles around.[9]

As previously noted, production of some precious metals and stones (gold, silver, and diamonds) is treated differently because the volumes involved are smaller and can be transported by plane or helicopter. Moreover, unlike the main nonferrous metals, which are exploited by large oligarchic groups, gold presents a peculiar situation. In addition to a few major players, there remains a tradition of small placer miners, who accounted for 38 percent of the 52 tons of gold produced by the peri-Arctic region of Magadan in 2021.[10] The application of sanctions has made them more vulnerable. Since the early 2000s, both small and large players have heavily relied on foreign equipment at all stages of production: extraction machines, conveyors, special excavators and bulldozers, and refining chains, which were more efficient than their Russian equivalents at the time.

Russian researcher Natalia Galtseva mentions the scramble for spare parts for this equipment, now under embargo or payable only in foreign currencies.[11] The oligarchs in these sectors have access to multiple channels for sourcing spare parts and equipment from the non-Western world, including orders from Russian suppliers in the Urals or imports from China, even if it means accepting decreases in productivity. Strategic sectors such as copper, nickel, and rare metals are also assured of receiving increased government aid, which is more challenging for small or medium-sized enterprises.

Hydrocarbons: The Breakthrough of Liquefied Natural Gas

The discovery of large hydrocarbon deposits on the Ob Plateau, starting in the late 1950s, significantly altered regional dynamics. Initially, Nikita Khrushchev's government doubted the extent of the reserves. Where oil was sought, gas was found, and more to the northwest than geologists had predicted. When developing hydrocarbon deposits in the Ob Basin the planners were faced with a problem. Many of the deposits were located in areas where the ground in summer was transformed into vast expanses of lake or marshland. To create roads or railroads that could be used in the warm season, one had to start by digging trenches in winter, removing the frozen surface soil down to the permafrost layer, then filling in the trench by building a kind of dike of dry materials, such as sand or gravel. The following summer, a real road or pipe could then be built on this embankment.

As a result, the beginnings of true industrial exploitation in the 1960s were very difficult because the railway, coming from the south, lagged behind the rapid multiplication of extraction sites. Very quickly, a decisive choice linked to the policy of détente between the two Cold War blocs was made, which would link Russia to its European clients for several decades. The chosen option was to build several oil and gas pipelines between the Ob deposits and the western border of the USSR: the Surgut-Polotsk Oil Pipeline in Belarus built in the late

1970s; the Urengoy-Uzhgorod Gas Pipeline in Ukraine inaugurated in 1984; and the Yamal-Europe Gas Pipeline, further north, through Belarus and Poland, inaugurated in 2006.[12]

The scheme was simple and relied on mutually advantageous long-term agreements despite persistent political differences. Soviet and later Russian decision-makers counted on multiyear delivery contracts that bound them to their European clients at negotiated prices and quantities. During the Cold War, this was not without tensions. The Soviets needed large-diameter pipes suitable for cold conditions and compression stations that they did not manufacture in adequate quantity or quality. On a couple of occasions (the SS-20 crisis in 1977, the Euromissile crisis in 1983), the West tried unsuccessfully to delay the construction of Soviet pipelines by blocking certain technology exports. Similar American and European pressures were exerted in the twenty-first century on new pipeline projects intended to export Ob gas to Germany. Nord Stream 1 was inaugurated in 2012, and Nord Stream 2 was completed in 2021 but never put into service due to the war in Ukraine. (It was damaged by an underwater attack in 2022 that seems to have been organized by the Ukrainian services.[13])

Russia's Crimea annexation, and active support for the secessionist movements in Donbas in 2014, triggered a new wave of sanctions. By blocking deliveries of deep sea drilling equipment, these sanctions aimed to impact the Russian oil sector, which had to cope with the depletion of the already exploited surface layers of the Ob Plateau deposits. However, the deeper levels (the Bazhenov Formation located between 2,700 and 3,100 meters below the surface, the Achimov at 2,500 to 3,200, and the Tyumen at 2,800 to 3,200) required the use of technologies that the Russian majors did not have in sufficient quantity, leading to a drop in production in the middle and northern Ob.[14] But the construction of Nord Stream 2 was launched in 2017—that is, after the annexation of Crimea and despite the first US sanctions—by a consortium of Gazprom, Royal Dutch Shell, E.ON, OMV, and Engie.

As early as 2007, a battle ensued over the Third Energy Package voted by the European Parliament two years later. Aimed at strengthening European supply security, it proposed the unbundling of production activities from those of transport and distribution.

Russia's Arctic

This directly targeted Gazprom's monopoly, which the EU sought to break.[15] The Russian authorities rejected the proposal. Paradoxically, by trying to force Gazprom to abandon control of the distribution or transport of gas to EU states, Brussels raised awareness of these issues on the Russian side, with an unexpected effect that would play a major role in the following years. The leaders of the giant Gazprom remained convinced that the pipelines connecting Russia to Europe were an absolute guarantee of stability and could not be challenged. Engaged in the construction of the two Nord Stream Pipelines and the Power of Siberia Pipeline to China, they neglected the innovations that were transforming the global gas market, namely the emergence of shale gas and the massive growth of liquefied natural gas, which had begun commercialization in the 1960s.[16]

New Russian players such as Novatek and Rosneft then emerged, challenging Gazprom's monopoly. Novatek is a private oil and gas company led by Leonid Mikhelson, who obtained gas exploitation licenses in the northern Ob Plateau and on the Yamal Peninsula, not without numerous upheavals. In 2009, Mikhelson partnered with Gennady Timchenko, the founder of Gunvor, one of the world's leading oil traders. This choice was decisive because it was Timchenko, reputedly close to Putin, who appears to have secured a strategic shift from the Russian authorities. As early as October 2010, the government supported the creation of the Yamal LNG project, led by Novatek, which involved gas extraction in northern Yamal, the simultaneous creation of a large gas liquefaction plant, and the port of Sabetta, connected to the Northern Sea Route and accessible year-round to mini icebreaker-class tankers.[17]

The extraordinary Sabetta project began in 2012 under unprecedented conditions, at 72° north latitude, on a flat and unprotected shore of the Ob Estuary, which was only accessible to navigation from June through August (see Figure 3.1). Ice storms necessitated the protection of the shore with special dikes to bring in construction materials because the project started in a tundra landscape practically devoid of any installations. After the construction of the piers, terminals, and the plant, special tankers, aided by icebreakers, can load LNG year-round.[18] For the realization

Figure 3.1 The LNG port of Sabetta on the Yamal Peninsula (MikhailSSShuterstock).

of this project, since the Russians were not proficient in liquefaction technologies, Novatek (holding 50.1 percent of the shares) opened the project's capital to foreign firms: French firm Total at 20 percent, China National Petroleum Corporation and the China National Offshore Oil Corporation at 10 percent each.

Additionally, the Sabetta port called upon multiple specialized contributors: American (Chicago Bridge & Iron Co., General Electric), German (BASF and Siemens), Japanese (JGC), and French (Technip and Vinci). The project immediately drew criticism from the United States, which was experiencing a boom in shale gas. The first sanctions were imposed in 2014, but the Chinese Silk Road Fund saved the project by purchasing a 9.9 percent stake and opening a line of credit.[19] On the Russian side, in addition to direct investments (the state took charge of building the port), Moscow granted unprecedented advantages for this project, according to *Forbes Russia*: a reduced VAT on equipment imports, exemption from property tax, and twelve years of exemption from export and extraction tax. Production began, and the first shipment of liquefied gas departed from Sabetta in December 2017, aboard Sovkomflot's special tanker, the *Christophe de Margerie* (named in honor of the Total CEO who died in a plane crash in Moscow in 2014).[20]

As early as 2013, Gazprom's export monopoly was lifted with Putin's announcement of the liberalization of this sector, giving Novatek and its LNG projects free rein.[21] That same year, other large-scale projects were envisioned within Russia, including with Western partnerships. An exploration and production-sharing agreement for gas fields in the Chukchi and Kara Seas was signed in front of Putin by Igor Sechin, the head of Rosneft, and Steve Greenlee, the head of ExxonMobil. The newspaper *Kommersant* even ran the headline: "Russian Continental Shelf Will Become American."[22] In 2019, Rosneft created a subsidiary, Vostok Oil, specifically to develop all its Arctic projects.

The Russian invasion of Ukraine shattered all these agreements. The decision by the European Union to boycott Russian hydrocarbons in April 2022 practically halted the entire pipeline network, which had been patiently established over the decades since the 1960s, including the most recent ones like Nord Stream 1 and 2. In practice, however, these sanctions have been largely circumvented. Regarding oil companies, the Russian majors redirected their deliveries to Asian clients by chartering a "shadow fleet" of tankers acquired on secondary markets—with serious risks for pollution given the old age of these tankers. Russia had invested nearly $10 billion to set up a ghost fleet of several hundred vessels, estimated at nearly 600 ships.[23] These have been said to carry 70 percent of Russia's seaborne oil product exports, and even 90 percent of its crude oil. The total transported volumes have quadrupled since April 2022.[24]

Russia's oil and gas sector has thus continued to develop, whatever the political and economic conditions. Since 2008, Lukoil, which holds a series of oil fields in the Nenets Autonomous District, has used an ice-free oil terminal in Varandey on the Barents Sea. This terminal allows the loading of mini-tankers (70,000 tons) that can reach Murmansk, where the crude is then transferred to giant tankers. Additionally, significant efforts are being made to develop production sites for refined products—these constitute an important and the most lucrative part of exports—in the Asian part of the country, closer to the new clients.[25]

As far as gas is concerned, Moscow is relying on the expansion of Novatek's projects. The private firm is undertaking the construction of several large LNG units on the Yamal Peninsula and the Gydan Peninsula (Arctic LNG 2 and Ob LNG, across the Ob Estuary from Sabetta), as well as LNG hubs in Murmansk and Vladivostok. In order to facilitate the realization of Novatek's LNG projects on the Yamal Peninsula, the government has launched a new shipyard in Belokamenka, on the northern coast of the Kola Peninsula, ready to fill orders from Russian Arctic gas players: here, in the peninsula's relatively favorable climatic conditions, the gas liquefaction units that will then be transported to Yamal, or to possible offshore platforms that are expected to multiply, are being built.

But as noted previously, Russia is also pursuing its policy of developing a network of major pipelines built in southern Siberia running toward China and the Pacific Ocean. Some projects, however, are being held back by hesitant Asian partners. Proposed by Moscow back in 2006, the Power of Siberia 2 Gas Pipeline through the Altai region and Mongolia remains at a standstill. Negotiations with Beijing are stalling over the purchase price for the gas proposed by the Chinese, and Putin's visit to Mongolia in August 2024 failed to produce immediate results even if the project remains, at least metaphorically speaking, in the pipeline.[26]

In response to the acceleration of this dynamic, Western pressures have intensified to block the implementation of these projects. The United States aims to deprive Moscow of this source of foreign currency that contributes to the war effort, but the reactions of European allies have been neither immediate nor uniform. For instance, with the tacit agreement of the French government, the French group Technip continued to deliver modules intended for Novatek's Arctic LNG 2 plant until the end of 2022.[27] For its part, it was not until December 2022 that the company Total announced its withdrawal from the board of directors of Novatek, while still remaining a shareholder and continuing to sell Russian LNG.[28] In June 2024, despite German reservations, the European Commission decided to include all Russian

LNG in the fourteenth package of sanctions to prevent any sale in the European Union.[29] These measures aim to prohibit the unloading of Russian LNG in European ports, to provide no technical assistance to ongoing projects, and to propose tracking Russian tankers transporting liquefied gas.[30] But at the end of 2024, around 15 percent of EU gas imports were still coming from the Russian Arctic, mostly through third-party sellers.[31]

On the Russian side, fingers are being pointed at the direct commercial interests of the United States, which sees an increased opportunity to sell its own LNG to Europe at a good price now that cooperation patterns with Russia are broken. The situation is uncertain because Novatek does not yet have its own fleet of specialized tankers, and the use of nuclear-powered icebreakers on the eastern part of the Northern Sea Route is very costly. Sovkomflot is awaiting deliveries of tankers from South Korean shipyards, who tend to follow US sanctions and therefore do not deliver to Russia anymore. Russian and Chinese shipyards could potentially replace the lost production from South Korea, but it is unclear when this might be possible.

A Race against Time for a New Regional Dynamic

In analyzing the changes that have occurred since 2000 in the Arctic Zone regarding the development of its natural resources and access routes to global markets, it is evident that a sort of race is underway among several major groups of actors with opposing interests.[32] Under Putin's leadership, the Russian state has positioned itself to regain control over its market for strategic raw materials and aims to accelerate their development, a process that began in the 1930s. To achieve this, it has managed to bring the oligarchic groups in line with its rules of play, which involve both developing these new factors of Russian power and ensuring exorbitant profits for the presidential clan and its associates. This top-down restructuring has not been without friction between the majors in the hydrocarbon and mineral sectors, which is undermining its coherence. In 2013 for instance, the federal agency managing mineral resources Rosnedr and the three companies Rosneft, Gazprom, and Lukoil competed for control of new deposits discovered throughout the Arctic Zone, from the Barents Sea to the East Siberian and Chukchi Seas. And the agreement between Nornickel and Russkaya Platina to set up a joint subsidiary in charge of Norilsk's new platinoid deposits was only signed in 2018 under direct pressure from Putin.[33]

Moreover, these developments occur within an administrative framework that has long been unclear. Moscow has been slow to clarify the agencies responsible for coordinating this strategic space. For years, under Yeltsin and then Putin, the approach has been to follow the Soviet practice of multiplying economic and social development plans[34] for various individual regions, while relying on the overall (including social) role of major enterprises. The attempt to merge several administrative subjects of the federation, launched in 2003 and which ended in failure in the Far North, illustrates the degree of the government's unpreparedness. The goal was to consolidate regions deemed too weak due to their demographics or economies by merging them with their neighbors. While the operation was completed in regions like Perm and Buryatia, it failed with the autonomous districts of Nenets and Yamalo-Nenets, which were much wealthier and more dynamic than their neighbors and categorically refused to comply with Moscow's directives. Due to their resistance, the merger project was abandoned.[35]

Some Russian experts have criticized the delay in drawing up a truly global strategy for the Russian Arctic: while Norway published its Arctic Strategy in 2006, Canada in 2009–10, Finland in 2010, Sweden and Denmark in 2011, Russia did not publish its own until 2012, and still singularly lacking in concrete aspects.[36] Furthermore, it was not until 2015 that a State Commission for Arctic Development was established. A sign of the government's hesitancy, it was placed in 2019 under the Ministry for the Development of the Far East, which subsequently became the Ministry for the Far East and Arctic Development.[37] A series of programmatic documents was then published. They outlined both the priorities and the types of aid and subsidies granted to various stakeholders in the region. The provisions established by the 2020 law "On State Aid for Entrepreneurial Activity in the Arctic Zone" concretized the government's intent to promote the growth of all economic activities in the Arctic Zone, whose scope has been expanded several times. These provisions primarily benefit the major Russian oil companies, but also include local small businesses and the traditional activities of the Indigenous peoples of the North, who are increasingly involved in the business of their regions.[38]

This new phase of industrial development involves a systematic reliance on automation and innovative techniques, both in industrial processes and in the format of installations, which are becoming more compact and lighter, such as platforms and new "shift cities." These setups make extensive use of robots and computer management. The first Russian oil platform, Prirazlomnaya, constructed in 2014 by Gazpromneft off the coast of Varandei in the Barents Sea (in Nenets district), serves as a model of technological integration. Built by the Ural plant Sevmash and resting on twenty meters of water, it combines drilling, extraction, storage, and tanker service. A clever system for managing crude oil and ballast within the platform allows it to withstand the movements of ice floes and storms without emitting pollutants.[39] It is the only platform operating on the Russian continental shelf; the joint project between Gazprom and Total, envisaged in 2007 for the Shtokman Gas Field in the Barents Sea, was abandoned in 2014 due to its unprofitability given its geographical location and the global rise of shale gas.[40] The Belokamenka LNG compression unit yard mentioned earlier is another example of these new approaches.

Another frequently cited example is the rich gold and silver deposits at Kupol, in northern Chukotka, discovered in 1995 and put into production since 2008. Developed by the regional authorities and a major Canadian investor, Kinross Gold, the deposits are accessible only by plane or via frozen water roads (*zimniki*); the workers, who are essentially industrial freelance contractors, rotate every few weeks in and out of an ultramodern shift city built to serve the site. Extraction is almost entirely automated, and the ore is semi-enriched on-site. However, the viability of this model and similar ones is questioned following the breakup with Western firms that had provided expertise and technology, and in 2022, the Canadians sold their shares to the Russian firm Highland Gold.[41]

Indeed, apart from the Prirazlomnaya platform, which was built primarily by Russian companies, most projects developed in the 2000s and 2010s in the Russian Arctic relied heavily on partners and imported technical solutions. From the perspective of Russian oligarchs, this approach offered significant time savings and the acquisition of essential experience in unfamiliar fields. Some in Russia regret that the state did not encourage these firms to sponsor research involving other Russian manufacturers of machinery and equipment. This indeed highlights one of the limitations of the country's new industrial strategy.

Questioning the causes of such inertia in Russia, Valerii Kryukov and his colleagues point for instance to a number of factors, including the weakness and inefficiency of state governance in the area, the unwillingness of companies involved in resource development to cooperate with each other and combine their efforts in terms of technical innovation, and the inability to analyze the positive experience accumulated abroad (particularly in Norway) in developing and producing the equipment needed for polar climate. They call for more systematic exchanges between extractive companies, specialized university research centers, and equipment manufacturing plants.[42]

Faced with the near monopoly over the raw materials and transport sectors in the economy of this unique region, various proposals for diversification continue to be supported by local authorities, although they remain marginal. One of the areas often mentioned in the Russian trade press is the need to develop local agricultural production despite the unfavorable natural conditions. The Arctic is indeed included in the scope of the 2016 law that offers one hectare of agricultural land to any volunteer homesteader in the Russian Far East. Initially launched only in the Primorsky Krai, this program proved effective enough to be extended first to the Far East and then to the entire Arctic Zone.[43] However, it must be borne in mind that soils in the Arctic and sub-Arctic are less developed, more skeletal, and fragile. The measure will therefore have somewhat less impact in the Arctic region itself. However, in Chukotka, for example, 87 percent of foodstuffs come from elsewhere in the Russian Far East or even Europe, and the cost of transporting them triples or quadruples the local price, thus encouraging local production experiments.[44]

As a result, small greenhouse complexes are springing up all over the place, made all the more profitable when they can draw on local energy sources such as thermal springs or cooling water from nearby factories. In the absence of natural humus, hydroponic farming techniques are being experimented with successfully. If subsidized by cities or regions, these experiments can reduce costs and provide some of the local population with fresh produce such as tomatoes, potatoes, and herbs.[45] However, despite climate change, the region's agricultural potential remains negligible, even in relatively more climate-friendly areas like Karelia. Due to skeletal soil and low light levels for much of the year, the share of agricultural land in the total area is particularly low: 0.19 percent in the Murmansk region, and just 0.15 percent in the Nenets okrug. It reaches 1.2 percent in Karelia and 1.8 percent in the Arkhangelsk region, but parts of these territories lie outside the official Arctic Zone.[46]

Fishing plays only a marginal role, apart from local fishing to feed coastal populations. In fact, in 2018, the European Union and nine states (Canada, the United States, Norway, Russia, Greenland, Iceland, China, Japan, and South Korea) declared a moratorium on industrial fishing over some 285,000 square kilometers (110,000 square miles) of the Arctic Ocean for a period of sixteen years, in order to restore a resource that had been severely weakened by the exploitation of hydrocarbons and other industrial activities.[47]

Without reverting to the Soviet practice of developing small light industry units everywhere, one can still notice a resurgence of projects aimed at reestablishing specialized equipment factories in the northern cities to avoid costly transportation from more central regions. These can include food production units from local livestock, adapted construction materials (such as the lumber industry), alternative energy production (such as wind turbines), and so on. However, the labor shortage, which is becoming crucial nationwide during the prolonged war period, is likely to severely limit these projects.

One of the fastest-growing new sectors is tourism, both on land and at sea. Hikers, wilderness campers (known as *dikie turisty*, "wild tourists"), and other trekking enthusiasts were already numerous in Soviet times. The relaxation of access conditions in the Far North, especially the lifting of certain restrictions in formerly "closed" towns and the multiplication of means of transport, has led to a relative boom in this type of tourism, which is still marginal yet nonetheless significant for regions with poor infrastructure. Several natural parks have recently been created, such as the Russian Arctic National Park in 2009 in the Arkhangelsk region. Comprising the northern part of the Novaya Zemlya Archipelago, it was expanded in 2016 to include Franz Josef Land and now covers a total area of 1,426,000 hectares (nearly 3,524,000 acres) of land, islands, and marine expanse. Many regions are also developing ski resorts, such as Bolshoy Budyavr on the Kola Peninsula.

As in Scandinavia, Greenland, and Northern Canada, tourists are attracted by the spectacular scenery and natural beauty, as well as by the opportunity to discover the way of life and traditions of the peoples of the north. Local authorities and residents soon found themselves confronted with the phenomenon of overtourism in particularly fragile natural environments, as well as criminal activities such as poaching or trafficking in mammoth bones and other protected local products or resources. Gradually, however, an active tourist industry has begun to emerge, with specialized agencies and new infrastructure aimed at a wide range of visitors, from the wealthiest (bear hunters transported by helicopter) to the most sporty. Starting with a handful of guidebooks published in the late nineteenth century, the number of books specializing in tourism in the Far North continues to multiply.[48]

Special mention must be made of Arctic cruises.[49] Since the 2000s, the Russian Geographical Society has been organizing the Barneo floating summer camp (on pack ice) on a former Soviet research platform of the same name, which welcomes around 250 researchers and tourists in April.[50] Specialized tour operators offer stays of varying lengths (several weeks if a major part of the Northern Sea Route is to be covered) on specially chartered vessels. A wealthy cohort (all-inclusive stays can range from $10,000 to $50,000) seeking unique experiences that combine navigation in these singular seas, the discovery of its fauna, and various activities on the pack ice make up a passionate clientele.

The war in Ukraine has all but put an end to tours that include passages through Russian territorial waters. The *Commandant Charcot*, a small luxury ice-class ship, currently only offers tours in the "Western" Arctic. For its maiden voyage in September 2021, taking advantage of an exceptionally warm and ice-free summer, it went as far as the North Pole. But the Russians are not to be outdone: for several years, a few icebreakers and freighters have been taking a limited number of Western tourists along the Northern Sea Route. In 2024, one of Russia's former nuclear icebreakers, the *Fifty Years of Victory* (launched in 1993), was converted into a cruise ship and now offers twelve- to thirty-day tours of the Arctic Ocean with a visit to Russian Arctic National Park. Similarly, the scientific exploration vessel *Professor Khromov* offers tours of the shores of the Chukotka River, Wrangel Island, and Beringia National Park,[51] and the Barneo floating station reopened to Western tourists in 2024.

As noted, the 2022 full-scale invasion of Ukraine led to the systematic imposition of sanctions and boycotts explicitly targeting Russia's Arctic projects, but this opposition predates the annexation of Crimea. The outcome of the battle to prevent Russia from exporting LNG produced by Novatek from the Arctic front may prove crucial, as it is one of the main new sources of revenue for the Russian budget. Since 2024, Western sanctions have also included the Russian diamond sector, primarily affecting the Republic of Sakha (Yakutia) and the Arkhangelsk region, as well as mineral exchanges on Western metal exchange markets. Sanctions target both commercial networks (such as transportation capabilities, financing, insurance of ships, pressure on new clients, and intermediary banks in these new markets) and technical support, with embargo measures on components necessary for the operation of highly automated equipment.[52]

On the Russian side, Moscow has so far managed to obtain equipment and components either through alternative foreign markets or by developing its own production capabilities. The key issue is whether the adaptation timelines are compatible with Russia's immediate war effort. In this particularly challenging situation, characterized by expanding Western sanctions and reduced Russian investment capacity, regional leaders and experts are seizing every opportunity to preserve the dynamism of the 2010s. One such hope is to align various Russian projects with the new "Polar Silk Road," one branch of the Chinese Belt and Road Initiative that passes through the Northern Sea Route.

In this spirit, Piliasov suggests making Norilsk a sort of coordination hub for the development of Russia's Eastern Arctic region. He envisions creating several research centers or think tanks there and organizing an annual forum similar to those in Saint Petersburg and Sochi, focused on Arctic development in collaboration with all interested Asian partners.[53] But his colleagues from Kola Research Center suggest that Vologda—a much more southern city—should take on the role of major regional coordinator for the entire Central Arctic.[54] The city has the advantage of being a good transportation hub but is located outside the Arctic Zone. Today, the creation of such a forum as that called for by Piliasov seems unlikely; however, as previously noted, Arctic projects take time to realize and should be assessed over the long term.

Despite the ongoing war effort in Ukraine and the various constraints—financial, technological, and commercial—that this conflict entails, Moscow continues to pursue its Arctic development strategy, a key element of its future growth. Many projects have been delayed, and some may not materialize for a long time, but this has been a characteristic of the Russian economy for decades and does not disrupt its long-term development momentum. However, beyond the waves of sanctions, whose effectiveness remains uncertain, the main obstacle to Russia's development strategy is likely to be demographic and social.

CHAPTER 4
WHAT DEVELOPMENT FOR THE FAR NORTH?

Russia's vast Arctic territory is composed of several subregions facing different development challenges. It can be divided into three major functional regions: (a) the Western Arctic (the Barents Sea and White Sea), where the largest portion of the population and economic activity is concentrated; (b) the Central Arctic (from the Yamal Peninsula to the Taimyr Peninsula), characterized by sparse settlements but intense extraction activity; and (c) the Eastern Arctic (from Sakha-Yakutia to Kamchatka), marked by sparse populations and isolated activity clusters. A second perspective is more administrative in nature: the Arctic of large cities with multinational urban populations (Murmansk, Norilsk, etc.) versus the Arctic of Indigenous peoples (the Republic of Sakha (Yakutia) and the autonomous districts of Nenets, Khanty-Mansi, and Chukotka). A third perspective is demographic: the growing Arctic (primarily the northern Ob River Basin and the Yamal Peninsula) versus the demographically declining Arctic (such as Chukotka). Each of these three categories approaches the challenges of regional development in the Russian Far North differently.[1]

The Far North in Russian Identity Construction

Within the circumpolar world, Russia has the longest history of conquering Arctic and sub-Arctic spaces. As early as the Middle Ages, the Novgorod Republic advanced into Karelia and toward the shores of the White Sea, trading with major Hanseatic cities of northern

Europe. The Pacific Ocean was reached in 1680 near Sakhalin, and Alaska came under Russian control in 1741. In the eighteenth century, Peter the Great financed major maritime expeditions to Kamchatka, including that of Danish Captain Bering, which led to the first mapping of the Arctic coasts.[2]

However, most of Russia's territorial advances have been made overland, from one river and its tributaries to another, rather than by sea. This is a central element of Russian Arctic identity. It is far more continental than maritime, and what is referred to in Russian as the Far North is the heart of this high-latitude continentality. Until recently, the term "Arctic" (*Arktika*) was reserved for the ocean itself, and it has been only with the internationalization of the region that the Russian authorities have begun using the term "Arctic" in its Western sense to describe both the lands beyond the Arctic Circle and the ocean together.

The Far North has long been part of the Russian national imagination. Since the nineteenth century, folklorists and nationalist thinkers have extolled traditions preserved from the Russian North (*russkii sever*), covering the regions of Arkhangelsk, Murmansk, and the Komi Republic (and also boasting the fact that they had never been invaded by their Swedish competitor during tsarist times or occupied by Hitler during the Second World War). In the early twentieth century, the tsarist elites were passionate about the European race to explore the poles, north and south, but the Romanovs did not invest in building an Arctic fleet. This would cost them dearly during the Russo-Japanese War of 1904–5, as the Russian fleet stationed in the Baltic Sea had to pass through the Suez Canal or around the Cape of Good Hope, rather than take the Northern Sea Route, in order to reach the theater of war in the Pacific.

Everything changed radically with the onset of the Soviet regime. As early as the 1920s, the Bolsheviks invested in the Northern Sea Route, launching several expeditions and a floating research institute to survey all the Siberian rivers and their access to the Arctic Ocean, while also innovating in the field of trans-Arctic aviation. The Stalinist decades accelerated this polar commitment. In 1932, a Soviet icebreaker became the first in the world to traverse the entire Northern

Sea Route in a single summer. Between 1934 and 1937, Soviet aviation set the world record for flying over the Pole from Moscow to the United States, entering into the world legend of aviation. The Northern Sea Route Committee (Komseveroput') accelerated the mastery of polar navigation to access mineral reserves. The Central Administration of the Northern Sea Route (Glavsevmorput'), described by historian John McCannon as "one of the greatest Soviet experiments in hypercentralization," controlled commerce, navigation, industrial and agricultural production, as well as population flows, before being dismantled in 1938 and replaced by several sectoral institutions.[3]

These years gave birth to the myth of the "Red Arctic," a frontier of Soviet civilization, the last virgin space on the planet that socialism was conquering through its industrial feats and its subjugation of nature.[4] During the Cold War years, the strategic position of the Arctic, not far from the US coasts, was emphasized, and the pioneering spirit remained very much present among the populations of the Far North's cities. Soviet urban planners experimented with polar utopias, designing cities that were adapted to their precarious environment, socialism, the culture of Indigenous peoples, and the so-called creative class.[5]

Since the 2010s, the region has been reappraised by the authorities as a major part of Russian national identity, where wilderness is to be preserved, and extreme conditions, human and industrial achievements, and international prestige are all intermingled. The Russian Geographical Society is leading this rediscovery in a way inspired by the US-based National Geographic Society. It funds scientific expeditions, memorialization (such as erecting plaques and Orthodox crosses, restoring abandoned sites, and documenting and archiving), and public marketing operations (including popular documentaries, photo exhibitions, contests, etc.), while also overseeing international polar cooperation.

Conservative and nationalist ideologues close to the Kremlin were among the first to interpret the restoration of the Russian presence in the Arctic as a form of "reparation" for the loss of post-Soviet territories, particularly Ukraine.[6] Alexander Prokhanov, one of the most prolific of these ideologues and leading the conservative think tank Izborsky Club, sees for instance a renewal of Russian messianism in what he calls "the Russian march toward the north" and the assertion that the whole Arctic Ocean constitutes Russian territorial waters.[7] Not without humor, he designates Gazprom as "the corporation of all the Russias" (on the model of the "Church of all the Russias") and notes that the Arctic is likely to become the source of both Russia's material and spiritual power, because "the Arctic civilization requires an incredible concentration of force in all domains. It will become, then, a sanctified 'common good,' in which the peoples of Russia will rediscover their unity, conceived by God as those to whom he destines great missions."[8] While the Russian government does not use such straightforward language, the idea of the Arctic as central for Russia's great-power remains very present. The peculiarities of Russia's Far North are also part of the idea of a national Arctic destiny.

Polar Urbanity: A Russian Peculiarity

The Russian Arctic is predominantly populated by settlers from other regions of Russia or the former Soviet Union living in urban environments.[9] It has the highest rate of circumpolar urbanization: between 66 and 92 percent of the population in the sixteen Arctic regions is considered urban, with the Murmansk region leading the way. Therefore, any discussion of demographic changes is intrinsically linked to issues of urban development, which is itself intrinsically connected to the boom-and-bust cycles associated with natural resource development.[10]

The two oldest cities of the Russian Arctic, Arkhangelsk and Yakutsk, were founded in the sixteenth and seventeenth centuries, respectively, the former as a port for the Russian Empire, and the latter as the center of the conquest of Eastern Siberia. Murmansk, on the other hand, was established just before the collapse of the tsarist regime. These large cities have managed to maintain multiple economic activities and significant cultural and educational influence. Murmansk and Arkhangelsk benefit from their port status, with major logistics and transportation companies and proximity to closed—meaning entry requires special permission—military cities housing personnel from

What Development for the Far North?

the Northern Fleet. Although outside the Arctic Zone proper, Yakutsk also has a diversified economic profile, ranging from gold and diamond extraction, the two main resources of the Sakha Republic, to agriculture in the south, transportation, and a vibrant tertiary sector encompassing small services, a cultural industry, and education.[11]

The cities that emerged during the Soviet era can be divided into two main categories. The first category, the pioneering cities, or frontier cities in the American sense of the frontier, was established in the 1930s, driven by vast mining projects managed by the Gulag administration. These cities have long preserved the hidden memory of the Gulag among their citizens.[12] These are often single-industry towns entirely dependent on a specific sector (such as Norilsk, Vorkuta, or Magadan, based on raw material extraction) or a mode of transportation (such as port activities for Dudinka on the Yenisei). In Norilsk for instance, between half and two-thirds of the population still works for the company Nornickel and its subsidiaries or affiliates, figures that are also found in other smaller towns dependent on Nornickel like Monchegorsk on the Kola Peninsula (see Figure 4.1).[13]

Figure 4.1 Norilsk city epitomizing Arctic urban infrastructure © Marlene Laruelle, 2015.

Russia's Arctic

The second category is what Nadezhda Zamiatina refers to as the "link city" (*gorod-zveno*), one that is intricately integrated into an urban network. This urban fabric ranges from administrative capital cities (with their political representation institutions, cultural and educational facilities, public and medical services, logistics, and transportation companies) to industrial cities specializing in a specific type of natural resource deposit and connected with "shift cities." These shift cities are located as close as possible to their resource deposits, with workers flown in by helicopter on rotations ranging from two weeks to a few months, depending on the sector and company. Notable examples of "link city" include Nadym, established in 1972; Novyi Urengoy, in 1980; Noyabrsk, in 1982; and Gubkinsky, in 1986.

This dense urban fabric, which is not found elsewhere in the circumpolar world, is the distinctive characteristic of the Russian Arctic, particularly in the three most developed peninsulas: Kola, Yamal, and Taimyr (see Table 4.1). Beyond the Arctic Circle, five major cities dominate: Arkhangelsk (296,000 inhabitants in 2024), Murmansk (266,000), Severodvinsk (155,000), Norilsk (176,000), and Novyi Urengoy (104,000). Yakutsk, with a population of 384,000, is also notable, and even if it was built just below the Arctic Circle, it is still constructed on permafrost and in extreme sub Arctic continental conditions.[14]

In some of these cities, wealth flows in abundance. In Western Siberia, the Khanty-Mansi district, known as Yugra for short in Russian, accounts for 60 percent of the economic output of Russian Arctic and sub-Arctic regions. Its two major cities, Khanty-Mansiysk (111,000 inhabitants) and Surgut (420,000), are part of this oil rent archipelago—boasting some of the highest GDP per capita figures in Russia.[15] At the other end of the socioeconomic scale are cities in full decline, such as Igarka on the Yenisei,[16] or small ports along the Northern Sea Route struggling to be revitalized, such as Dikson, Tiksi, or Pevek in Chukotka, with only a few thousand inhabitants each. This diversity of local socioeconomic conditions has a major impact on the strategies and planning of municipalities in terms of their socioeconomic development opportunities and adaptability to climate change.[17]

What Development for the Far North?

Table 4.1 Major Cities of the Russian Arctic by Population, 1990–2024

City	1990	2000	2024	% change from 1990 to 2024
Anadyr	17	11	13	−23.5
Apatity	88	69	52	−40.9
Arkhangelsk	391	361	296	−24.2
Dikson	4.4	1.4	0.3	−93.1
Dudinka	32	26	19	−40.6
Igarka	18	9	3.5	−80.5
Murmansk	442	376	266	−39.8
Nadym	52	44	44	−15.3
Naryan Mar	20	18	24	20
Norilsk*	178	140	176	−1.1
Novyi Urengoy	93	89	104	11.8
Pevek	12	6	4	−66.6
Salekhard	31	33	51	64.5
Severodvinsk	248	220	155	−37.5
Tiksi	11	6	5.8	−47.2
Vorkuta	116	89	56	−51.7
Yakutsk**	192	195	384	100

Notes: *It should be noted that, throughout the Soviet period, the city of Norilsk was not included in statistical yearbooks in the north of Krasnoyarsk krai but in the south, leading to frequent confusion.

**Yakutsk does not belong to the Russian Arctic Zone administrative delimitation.

Source: Compilation by the authors based on Rosstat and Russian media.

Russia's Arctic

With the onset of Western sanctions, this regional diversity has increased. Some cities, like Naryan-Mar, the small capital of the Nenets Autonomous District, have suffered from sanctions that slow down the development of the Indiga port, designed for cargo transshipment and LNG from adjacent fields. Other regions like Arkhangelsk and Murmansk, with their military-industrial complexes and shipyards, have benefited from new public investments related to the war in Ukraine.[18] Politically, diversity is also evident: Yakutsk, for example, presents a more democratic outlook compared to cities controlled by major firms, like Norilsk, which is managed by Nornickel; or Salekhard, managed by Gazprom. Arkhangelsk, for its part, experienced large-scale environmental protest movements from 2019 to 2020 against the construction of a massive landfill, with protesters creating a form of local resistance zone.[19]

In spite of—or perhaps because of—the daunting material conditions, the inhabitants of polar cities have developed a strong sense of local patriotism. They refer to central Russia, from which they feel distant, as "the continent" (*materik*) or "the great land" (*bol'shaia zemlia*). The memory of Soviet generations who built these exceptional urban spaces from scratch, even amidst pollution and environmental destruction, plays an important role in local identity, as does the social capital accumulated over several decades.[20] For example, numerous studies have shown that a significant number of residents of Chukotka were not interested in being relocated to southern Russia, even after retirement, because they had built strong social capital in the north.[21] This sense of belonging persisted in the 2000s and 2010s, despite increased mobility among young Arctic citizens, both within the Arctic Zone and between the "continent" and the Far North.[22]

Demographic Flux and Reflux

The Arctic urban fabric has experienced sudden demographic transformations due to changes in population mobility. With the collapse of Soviet industries and the abrupt cessation of financial and material benefits provided by the state to those working in the

What Development for the Far North?

Far North, one-sixth of the population left the region in only one decade, between 1989 and the early 2000s. Over the same period, the Magadan and Chukotka regions lost between half and two-thirds of their populations, while Taimyr and Yamalo-Nenets lost a quarter. Several dozen small villages were declared "uninhabited" in the 2002 and 2010 censuses. Larger cities like Murmansk and Arkhangelsk faced a more gradual depopulation (with population losses of 40 and 25 percent, respectively, over three decades), exacerbated by the proximity of major metropolises such as Moscow, Saint Petersburg, and Yekaterinburg, which have attracted youth and former Soviet cadres.[23]

In the 1990s, everyone who could find a job outside the Far North left, leaving the poorest individuals, often the elderly, as well as Indigenous populations, less mobile, in ghost towns. This phenomenon created what are known as "poverty holes": real estate prices collapsed so much that residents of these deserted towns had no hope of selling their properties to buy elsewhere in the country. Vorkuta, a former Gulag town specializing in coal extraction, lost almost two-thirds of its population since the early 1980s, declining from 200,000 inhabitants (with its suburbs) at the last Soviet census of 1989 to only 67,000 in 2024. The small worker towns surrounding the city have been abandoned, reclaimed by nature, while the city itself has become an archipelago: entire neighborhoods, once embodying the spirit of an advanced socialist outpost in the Arctic, have been completely emptied of their inhabitants, with those who remain retreating to the city center with its services.[24]

However, since the 2000s, and more so in the 2010s, there has been a remigration to the Arctic and, consequently, a growing diversification of the urban landscape, with cities such as Salekhard, Novyi Urengoy, Naryan-Mar, and, further south, Khanty-Mansiysk experiencing a demographic boom of 15 to 30 percent. Today, three Arctic regions are experiencing a steady population increase: the Autonomous Districts of Khanty-Mansi, Nenets, and Chukotka, the first mainly due to the arrival of external workers, the second due to a mix of external arrivals and natural growth, and the last primarily due to natural growth.

Russia's Arctic

While the Republic of Sakha (Yakutia) is experiencing a gradual decline, its capital, Yakutsk, is undergoing a demographic explosion: its population has grown from 186,000 in 1989 to 384,000 in 2024. This unique demographic dynamism is founded on the mass exodus of the ethnic Yakut population from rural parts of the republic to the capital city in a form of "ethnic replacement," as Russians and other Slavic-speaking populations leave for the European parts of the country. Unlike Sakha, the regions of Murmansk, Arkhangelsk, Komi, and Karelia, on the other hand, are facing both migration outflows as well as a decrease in natural growth, trends that are expected to intensify in the coming years.[25]

Russia's Arctic thus displays both shrinking and growing cities, sometimes near each other, as in the case of Vorkuta and Salekhard, located on opposite sides of the Ural Mountains and separated by just 150 kilometers (93 miles). While Vorkuta is dying demographically, Salekhard, often presented as the Gazprom capital, is one of the most rapidly developing (and one of the wealthiest) urban centers in Russia, growing from 16,000 in the 1960s to more than 50,000 inhabitants nowadays.[26]

In general, the population of the Russian Arctic is younger than the national average (in 2019, those under fifteen years old represented 21 percent of the Arctic regions, compared with 18.7 percent nationally) and, in dynamic regions, more educated than the national average. Indeed, major energy projects, particularly in Yugra, attract a young, skilled workforce seeking professional opportunities, and Indigenous peoples have higher birth rates, especially in the Sakha Republic and the Nenets Autonomous District. Retirees are, as one might expect, fewer in number than the national average, but their numbers are increasing, since more and more people are staying in the Arctic even after they retire. However, as during the Soviet period, life expectancy is lower in the Arctic, even if data are difficult to gather, as many people move to Russia's warmer regions after retirement. The Arctic also has a different male-to-female ratio compared to the rest of the country, with prevailing male-dominated professions.[27]

In terms of migration, the Arctic has been one of the most dynamic regions in Russia these last two decades: many people are leaving, but many are arriving too (see Tables 4.2 and 4.3). Young

Table 4.2 Population of Russian Arctic Regions on January 1 (in Thousands, 1990–2020)

City	1990	2000	2010	2020	% change from 1990 to 2020
Arkhangelsk Region	856	745	696	665	−22.3
Chukotka Autonomous Okrug	164	61	51	49.5	−69.8
Dolgano-Nenets Autonomous Okrug, including Norilsk	365	299	245	243	−33.4
Murmansk Region	1165	941	800	732.9	−37
Nenets Autonomous Okrug	54	41	42	44.4	−17.7
Republic of Karelia	179	156	130	111	−37.9
Republic of Komi	342	251	191	154	−54.9
Republic of Sakha (Yakutia)	149	85	73	67	−55
Yamalo-Nenets Autonomous Okrug	495	496	524	547	10.5
Total Russian Arctic Zone	**3,769**	**3,075**	**2,752**	**2,614**	**−30.6**
Khanty-Mansi Autonomous District*	1,282	1,360	1,521	1,687.7	31.6
Magadan Region*	392	202	159	139	−64.5

Notes: * Although northern, the Magadan region and the Khanty-Mansi district are not officially part of the Russian Arctic Zone. Data include only the districts that do belong to the Arctic Zone, not necessarily the whole administrative entity.

Source: Compiled by the authors, based on data from Rosstat, rosstat.gov.ru.

Table 4.3 Migration Flows by Region (Number of Individuals) in 2020

	From Russia	From abroad	Total
Arkhangelsk Region	−2,455	+401	−2,054
Chukotka Autonomous District	−892	+132	−760
Khanty-Mansi Autonomous District	−567	+6,066	+5,499
Komi Republic	−4,665	+1,341	−3,324
Krasnoyarsk Krai*	−1,445	+3,287	+1,842
Magadan Region	−944	+325	−619
Murmansk Region	−3,993	−466	−4,459
Sakha Republic (Yakutia)	−2,632	+8,697	+6,065
Yamalo-Nenets Autonomous District	−938	−133	−1,071
Total for the Russian Arctic Zone	**−18,531**	**+19,650**	**+1,119**

Note: * Part of the Krasnoyarsk Krai is not in the Arctic Zone.

Source: Compiled by authors based on data from Rosstat, rosstat.gov.ru.

people from the North Caucasus, particularly from Dagestan, or from other Russian rural areas have moved to Yugra, followed by labor migrants from Central Asia and the South Caucasus, drawn by higher wages and the need for workers who can develop the service sector. This is reflected in Table 4.3, where, at least for the year 2020, departures to other Russian regions were fully offset by arrivals from abroad. As Nadezhda Zamiatina and Ruslan Goncharov explain, "contrary to the widespread stereotype that these migrations are based on the return of people who migrated to the North and Arctic during the Soviet era, there is a two-way migration."[28] These flows are often based on a pairing of regions, for instance between Murmansk and Novgorod, Madagan and Belgorod, Kamchatka and Kaliningrad, the Yamal-Nenets Autonomous District and the

Republic of Bashkortostan, and so on, which can be explained by historic and sociocultural patterns.²⁹

Arctic Multinationality and Polar Islam

This attractiveness has enhanced the multinational nature of major Russian Arctic cities. This is not new. The possibility of upward social mobility promised by the Soviet regime's industrialization and urbanization plans pushed people to move toward the pioneering new fronts of the Far North. In the 1960s, Azerbaijani petroleum engineers became the first to take advantage of their long-standing oil knowledge to move to Western Siberia and contribute to the exploitation of new oil fields on the Ob River, such as Samotlor, and the discovery of new gas fields further north in the Yamalo-Nenets Autonomous District, such as Medvezhye.³⁰ They were followed by Tatars and Bashkirs, who likewise invested in Siberia's growing energy sector. This first generation of skilled workers is revered as the "founding fathers" of local Muslim communities, even if their religious identity was nonexistent or underground during the Soviet decades.

One can schematically identify four main waves of migration that have shaped the Muslim landscape of Russia's Arctic cities.³¹ First, in the 1980s, at a time when the Soviet economy was slowing down and ethnic tensions were emerging in the southern republics, some young men finishing their military service in Arctic cities decided not to return home. This is mostly the case of young Azerbaijani men who, after serving in the Northern Fleet or somewhere else in the Far North, chose to stay there and build a life for themselves. They were the first entrepreneurial generation, which slowly took shape in the final years of the Soviet Union, and are now often considered "notables," well established in their respective cities and helping new migrants to settle there.

A second wave arrived in the early 1990s: some were explicitly fleeing conflict in the South Caucasus or the civil war in Tajikistan (1992-7); others were just looking for job opportunities. In the years that followed the Soviet regime's collapse, Russia's Arctic cities, with their still-functioning extraction industries, looked like islands of stability and relative prosperity in an otherwise bleak context. Miners, from coal cities in Kyrgyzstan or Uzbekistan, for instance from Kyzyl-Kiya or Angren, whose mines had stopped working almost immediately after the collapse of the Soviet Union, decided to move to Vorkuta, which was still operating mining.

The third and largest wave took shape in the 2000s. In the first half of the decade, Russia's booming economy began attracting labor migrants from the whole post-Soviet space, a phenomenon that reached an unprecedented level in the second half of the decade. In 2013 for instance, the UN calculated that there were approximately 11 million foreign migrants in Russia, making it second only to the United States.[32] Russian citizens who had stayed in Arctic extraction cities during the 1990s because they offered social and financial stability began leaving for better-located cities in the 2000s with the new economic growth. The rise of the service sector improved the quality of life in Russia's European regions, rendering large and medium-sized industrial Arctic cities less attractive. Consequently, fewer qualified Russian citizens moved to the Far North to take up the offers made by the major companies there, which then looked to bring in foreign migrants en masse, for whom a good salary was worth living in harsh climatic conditions.

This replacement of Russian labor with immigrant workers has been visible in several industrial sectors—mining in Vorkuta and Talnakh, fishing in Murmansk, and smelting plants in Norilsk. These long-distance commuters (LDCs) represent a specific case in Russia's industrial workforce. Shift work (*vakhtovyi metod*, short-term tours of duty on extraction sites from a base city) is considered to be at the bottom of this industrial social ladder due to its tough working conditions that entail health risks and spending an extended period away from one's family. LDC jobs are increasingly performed by labor migrants: in the small village of Koachovo, near Apatity, for instance, several barracks are receiving Uzbek and Tajik migrants for mining jobs in the Khibiny Mountains.[33]

Yet the majority of labor migrants who came in the 2000s have secured jobs in niches outside the industrial sector, in the booming service industry. They occupy several specific niches: market stands selling fresh fruits and vegetables imported from their home countries; flea markets; networks of cafes, restaurants, supermarket chains, and nightclubs; small construction and transportation companies (taxi and bus drivers); and the whole car repair business. This boom in services deeply impacted the geography of Russian Arctic (and not only) cities: new construction changed the faces of downtowns and Soviet-era suburbs alike, with the emergence of ethnic neighborhoods and a symbolic hierarchy between districts based on the quality of services offered in terms of small businesses, leisure, and consumption.

The fourth wave began at the end of the 2000s and continued into the 2010s. It includes two groups of migrants: second-generation Central Asians (whose parents had already migrated to somewhere in Russia) and internal migrants from the North Caucasus, mostly from Dagestan. Although some of this movement may initially have been related to political tensions in the North Caucasus following the two wars in Chechnya, the flows are now linked to economic conditions and the job market. Between 5 and 6 million Russian citizens move from one Russian region to another one every year, a large proportion of whom are North Caucasians, particularly from Dagestan, which has a negative net migration flow of tens of thousands of people per year.[34] This trend has intensified to the point where, today, Yugra has become the third-largest region of Islamic presence in Russia, after its two historical centers, the North Caucasus and the Volga-Ural region.[35]

Muslims (as understood in terms of cultural affiliation, not necessarily by religious observance) now form an important part of the Arctic urban fabric, with their own neighborhoods, a distribution of occupations by ethnic group, and mosques that have been built either with community funds or with financial support from major companies like Gazprom. For example, Norilsk prides itself on hosting the northernmost mosque in the world.

Whereas the Russian Orthodox Church has been gaining in symbolic power in the Russian public space without seeing an increase in everyday religious practice, Islam is experiencing a genuine revival both in the public space *and* in the practice of its believers. This sociological trend is even more visible in terms of migration: Islamic belonging, more than ethnic identity, has progressively become a marker of identity among many Caucasian and Central Asian migrants in Russia. Mosque communities have grown in influence by becoming places to socialize, while references to a halal way of life have sprung up. Low status and the feeling of insecurity in a Russian environment are compensated for with Islamic piety. Migrants who die in Russia during migration are increasingly considered martyrs (*shahid*). As migrants occupy the lowest rung on the ladder of Russia's ethnic hierarchy, they turn to Islam to exhibit religious maturity and thereby gain social respect.[36]

This polar Islam is multinational, Russian-speaking, and well acclimated to the extreme conditions of the Far North.[37] These migrants, whether from within the Russian Federation or from Central Asian and Caucasian republics, have embraced the pioneering spirit of polar cities: their Arctic migration is a symbol of social success and entrepreneurship in their countries of origin, where marrying a "man of the North" (*severianin*)—someone who has succeeded in the Far North—is highly valued in the regions of origin.[38] However, the Russian polar regions continue to face a shortage of labor, particularly qualified personnel, a situation exacerbated by the war in Ukraine. For example, as stated by Florian Vidal, "between 2022 and 2023, the regions of Arkhangelsk and Murmansk, and the republics of Komi and Karelia, saw their labor forces decrease by 60,000, 40,000, 35,000, and 31,000 people, respectively."[39] The reliance on foreign migrants is not enough to fill this gap, especially since, as in the rest of Russia, it provokes mixed feelings among the local population and is now severely restricted by the authorities.

Indigenous Peoples and the Difficult Assertion of Native Identities

There is another form of multinationality that is rarely recognized in the Russian context, that of Indigenous nations. Russia is unique in the circumpolar landscape in that Indigenous communities constitute only a small percentage of its Arctic population. Whereas they represent 80 percent of Greenland's population, 50 percent of Canada's, 20 percent of Alaska's, and 15 percent of Norway's Arctic regions, they make up less than 5 percent of the population of Arctic Russia.[40]

Although Indigenous peoples have a more solid demography than Russians and have therefore seen their share of the Arctic population slowly increase over the past two decades, their rights remain fragile. Moscow does not consider the Arctic to have a specific status due to the presence of Indigenous peoples, and its reading of the region is still very much shaped by the colonial past, the memory of an easy conquest of territories deemed unpopulated, and the exploitation of the region's subsoil resources.

In Russian, the conquest is described as *osvoenie*, meaning "to make one's own," a positive term with strong emotional connotations, distinct from terms like colonialism or imperialism, which are rarely used in Russian historiography to describe the conquest of Siberia or the Arctic.[41] Indigenous populations were absorbed in various ways, either through peaceful assimilation or military force, and were integrated into the symbolic hierarchy of tsarist Russia, which, like the rest of Europe at that time, claimed to bring civilization to backward peoples. The Soviet Union partially changed this dynamic by recognizing cultural rights for Indigenous peoples, particularly those referred to as the "small peoples of the North" (*malye narody Severa*: those whose languages had fewer than 50,000 speakers), including, in theory, the right to preserve a traditional way of life.

However, the practical implementation of these rights was adapted to the political context of authoritarian socialism. Indigenous peoples were both collectivized and sedentarized, linguistically and culturally Russified (children were sent en masse to boarding schools and lost connection with their ethnic background, with the exception of some elite schools), while benefiting from administrative mechanisms that promoted their Sovietized culture and the formation of a national elite.[42]

These legacies are still palpable in contemporary Russia.[43] Although it is a federation, Russia is not an ethno-federation of regions equal in rights, and the hypercentralization under Vladimir Putin has further reduced the rights of national republics as well as Russian regions. In the Arctic, only one national republic stands out in terms of international recognition and cultural autonomy: the Republic of Sakha (Yakutia). The local government has invested in teaching the Sakha language and has indigenized the education system by creating nomadic schools in rural areas.[44] The Komi and Karelia republics do not display such strong identity, with their titular groups representing only a small minority (a quarter and just 5 percent, respectively) of their populations. Other administrative entities, such as Chukotka, Yamalo-Nenets, Nenets, Khanty-Mansi, and so on Autonomous Districts, have reduced cultural autonomy.

The notion of "small peoples of the North" used during the Soviet era has been contested, perceived as negative, and today the federal administration simply speaks of the "Peoples of the North" to refer to forty Indigenous nations. Currently, there are about 45,000 Nenets, 37,000 Evenks, and 31,000 Khantys, with numbers falling sharply for other groups like the 16,000 Chukchis or 12,000 Mansi.[45] Other ethnolinguistic groups are made up of members who number only in the thousands, hundreds, or even dozens of speakers. Furthermore, the status of "People of the North" only applies to communities living on their traditional lands, while a growing number of Indigenous peoples are settling in urban areas. It does not include the Sakhas, Komi, and Karelians, who are too numerous for this legal definition, although of course they consider themselves Indigenous.

The Sakhas, the largest group, with 460,000 members, have long protested against being regarded merely as "settlers" who came from Southern Siberia. Since 2016, the Constitution of the Republic of Sakha recognizes the Sakhas as Indigenous peoples, while also acknowledging specific rights for other less numerous Indigenous groups living within the republic (Evenks, Evens, Dolgans, etc.). The republic's leaders have tried to preserve the languages of minority peoples within the official school framework, but what can be done to preserve the language of a people numbering fewer than a thousand

What Development for the Far North?

individuals when the last native speaker (i.e., having learned his language in the family home) has disappeared?

Russia has not ratified the 1989 UN Convention on Indigenous and Tribal Peoples (ILO C169), as it would entail difficult negotiations on landownership issues, a crucial topic for Moscow given the strategic nature of Russia's energy industry. However, the authorities, who had long refused to consult local communities on subsoil use, have been forced to partially, though still modestly, take into consideration these issues.[46] In 2009, they established the first register of traditional use territories and provided symbolic compensation to communities that lost access to these lands due to the activities of large extraction companies.[47] The Nenets have been the most active in denouncing the theft of their ancestral lands by mining and energy corporations. They have organized numerous protests in the Khanty-Mansi and Yamalo-Nenets regions, complaining about the loss of their reindeer herding lands and the environmental degradation.[48]

For several years now, any new license for exploration and extraction of mineral resources in the Khanty-Mansi region has had to obtain the agreement of Indigenous communities. These agreements include compensation for the loss of land, commitments to invest in local infrastructure, and the hiring of Indigenous peoples. In practice, communities cannot refuse new developments and can only negotiate better financial or material conditions. In the Republic of Sakha (Yakutia), the presence of the national diamond corporation, Alrosa, and several smaller gold mining companies has also led the local government to pass a unique law in Russia. This law requires companies to undergo an ethnological audit to assess the societal costs borne by Indigenous communities when a new license is granted.[49]

Like other sectors of Russian civil society, Indigenous activism has been gradually restricted by the authorities. The association representing the Indigenous peoples of the North, RAIPON, established in 1990, still includes forty nations, representing a total of over 250,000 people. It was particularly active in the 1990s and 2000s, supporting Indigenous protest movements against major energy corporations, increasing its international visibility, and building circumpolar contacts. This led to difficulties when it began to be scrutinized by the Russian administration for its links with its Western counterparts, being seen as a potential fifth column.[50] RAIPON was briefly closed in 2012, and then reopened under new, more conciliatory leadership the following year. Since then, it has been regularly accused by the authorities of either corruption or having too close ties with foreign entities, and its role in the Arctic Council has been a point of tension between Russia and other member states.[51]

Over the years, many environmental protests have taken shape among Indigenous peoples—for instance, Shor people protesting the mining of coal in the southern Kemerovo region of Siberia, or Izhma Komi people contesting the extraction of oil in the northern Komi Republic. Indeed, mineral extraction is posing not only structural threats in terms of economic rights, health, and the environment but also cultural threats to Indigenous livelihoods and ways of life.[52] In the first months of the war in Ukraine, many young men from Indigenous peoples were recruited to the front, in a higher proportion than average Russians, which also sparked waves of protest in Sakha-Yakutia and more southern republics.[53] This recruitment is a typical case of the intersection of class and race in the Russian context, where ethnic republics and regions display a higher rate of young, unemployed, rural men forming a kind of captured clientele for military conscription.[54]

However, administrative divisions and legal statuses are only part of the challenges faced by Indigenous peoples. Their demographics have been very fragile for decades, with higher mortality rates than the Russian average, especially among men; widespread social issues such as alcoholism; and still significant infant mortality. These realities are not unique to Russia and are seen among Indigenous nations in North America, particularly in the United States.[55] Knowledge of native languages is also declining in favor of learning Russian (seen as a necessary tool for social advancement), although it is experiencing a symbolic revival among larger Indigenous groups like the Sakha. Common in tsarist Russia, religious conversions have reappeared after the collapse of the atheist Soviet regime, with some Russian Orthodox, as well as Protestant, groups proselytizing among Indigenous communities, which continue to practice shamanism.[56]

A massive urbanization movement is also underway, with growing number of young Indigenous peoples choosing to leave their villages and move to cities—a phenomenon that is not unique to Russia but is also seen across the circumpolar region. For a long time, the city symbolized colonial culture, a place where intergenerational ties were lost due to the system of boarding schools for Indigenous children. However, the difficulties in maintaining a traditional lifestyle and the appeal of educational and cultural opportunities offered by cities have shifted the dynamics, and increasing numbers of young Indigenous peoples now wish to reinvent themselves in an urban environment.[57]

Urbanization dramatically transforms Indigenous identity, disconnecting it from the traditional subsistence economy. This is usually accompanied by a loss of the native language and traditional ecological knowledge, along with deep transformations of kinship identity. Until recently, studies of Indigenous peoples in urban conditions focused primarily on all the social illnesses linked to difficulties to integrate into the urban fabric—such as social marginalization, unemployment, and suicide and alcoholism rates—and physical and symbolic dispossession rather than on the creation of new meanings. Since the second half of the 2000s, a more diversified perspective has emerged, making it possible to identify how indigeneity can be mobilized in urban politics and confirming an Indigenous agency that cuts against perceptions of Indigenous peoples as being exclusively victims.[58]

Some Indigenous groups, such as the Nenets, remain heavily rural, while others—such as the Mansy, Komi, and Sámi—are more urbanized. But even Nenets are progressively joining the trend toward urbanization: the Nenets population has doubled in the past eighty years, but only 1,000 more people are living in the tundra, where Nenets have traditionally engaged in reindeer-related activities, than were resident there at the time of the 1939 census. Precise figures may vary dramatically depending on the region, but the general trend is that of a rapid urbanization of Indigenous peoples: in the Magadan oblast, for example, urban dwellers grew from 33 percent of the Indigenous population in 1979 to 69 percent in 2010. In the cities of Khanty-Mansiysk and Salekhard, Indigenous peoples represent about

Russia's Arctic

20 percent of the population. They make up more than 50 percent of the residents of Yakutsk, the highly dynamic capital of Sakha, and over two-thirds of the 20,000 inhabitants of Dudinka, the port city on the Yenisei River from where Nornickel's icebreaker tankers depart for international markets. In all cities where Indigenous peoples form a significant community, a local identity is being reinvented, capable of combining urbanity with indigenous heritage, such as maintaining contact with traditional lifestyles (hunting, fishing, reindeer herding, etc.), cultural practices, rituals, and cuisine from Indigenous cultures, as well as craftsmanship.[59]

Yakutsk has now become the true Indigenous capital of the Russian Arctic. It boasts a growing Indigenous middle class, a dense network of cultural institutions, and native higher education establishments such as the Institute of Languages and Cultures of the Peoples of the Northeast (ILCPN) at the North-Eastern Federal University (NEFU) and the Arctic State Institute of Culture and Arts (AGIKI). The city is seeing a rise in energetic ethnic entrepreneurs who are creating small businesses while cultivating an Indigenous image in areas such as urban planning, tourism, and international relations.[60] As the leading Indigenous city of Russia's Arctic, Yakutsk paves the way for imagining the form Indigenous urban agency could take in a transformed Russia.

CONCLUSION

The Arctic stands at a crossroads: it could become embroiled in the current geopolitical escalation, evolving into a new point of tension between Russia and the West, akin to an extension of the Baltic zone. Alternatively, it could be partially saved from this conflictual situation and revert to its historical role as a space where littoral countries make attempts, however modestly, at forms of dialogue. Among Arctic stakeholders, there have been calls to revive scientific cooperation, particularly on climate issues and polar science, as well as cross-border and person-to-person relations. This is essential to prevent erasing over thirty years of Arctic dialogue-building, as well as to keep up an international cooperation on global issues that bypass political and geopolitical divisions, such as with climate change and biodiversity preservation.[1]

Actors from the Global South also wish to keep the Arctic free from Russia-West tensions, which they view as regional rather than global. They see the Arctic as a global commons that everyone is concerned about preserving. Moscow's intent to involve the Global South in the Arctic may therefore support a globalization of the region, which could potentially slow geopolitical fallout from it. However, it could also create other forms of tension in the medium and long term, such as around Chinese presence or the increasing commercialization of coveted Arctic resources. While one often thinks of the legal disputes over the continental shelf and its minerals, other resources such as fish stocks and possibly freshwater (to be desalinized to become drinkable) could find themselves at stake too.

What is certain is that the internationalization of Arctic affairs and the status of the Arctic Ocean more broadly (though not its coastlines, which belong to their respective nations) as a global commons will grow in the coming years. For Western countries, the strategic intertwining of Russia and China and the potential linkage between the Arctic and the Indo-Pacific region—the latter already a major center of geopolitical tensions in the twenty-first century—will remain sources of concern. The Trump's administration's renewed interest into controlling Greenland and facing Chinese presence in the Arctic in a more confrontational way confirm that the region will remain one of the centers of great powers' projection.

While Russia's conquest of its Arctic territories is the oldest and a quite unique phenomenon among the circumpolar regions, the exploitation of natural resources and attendant urbanization are not only Soviet but also Western legacy of a twentieth century marked by intensive resource extraction. This heritage of "petromodernity" continues to heavily influence the Russian perspective on the region: the Arctic is predominantly seen as an economic territory to be exploited, a treasure trove of raw materials that significantly contributes to the state budget, and a strategic space upon which national sovereignty is projected. While the Arctic's role in the planet's environmental balance is not entirely disregarded by the Russian authorities, it remains subordinated to strategic and economic considerations. The crucial role of the taiga and tundra in preserving the planetary biosphere, as well as the respect for Indigenous peoples' rights, are perceived as secondary.

The effects of climate change are expected to intensify: the Russian Arctic is the most affected region globally in terms of rising temperatures, leading to profound and rapid changes in both terrestrial and maritime environments. While Russian authorities might view some transformations as positive, the growing climatic imbalances are likely to result in an increasing frequency of wildfires in the taiga; new invasive species and public health impacts; coastal erosion; and, notably, accelerated permafrost thaw affecting economic infrastructure and settlement areas. Climate change will certainly compel all Arctic stakeholders to reimagine the future of the region and the livability of its cities.[2] Russia's Arctic will thus continue to grow as an archipelago, a vast, natural, and largely uninhabited landmass, dotted with urban centers like man-made islands, facing three key issues: connecting these islands to the mainland, finding a way for Indigenous cultures

Conclusion

and knowledge to survive, and creating a new urban culture that can live in harmony with its environment.[3]

Beyond the impacts of climate change, the economic development of these northern territories faces significant short-term challenges. Their medium- and long-term sustainability remains in question. Even if the current massive investments into the military-industrial complex have revitalized some Arctic and sub-Arctic regions, this still does not mean that Moscow has the means to sustain expansive urban infrastructure in the Far North on such a scale for decades to come. The securitization of the Arctic is, in itself, a guarantee of Moscow's continued political will to invest in the region. But money is not the only issue: human capital is as much, if not more, pressing. Indeed, regardless of the level of automation—which is high in new hydrocarbon extraction units—workers will still be needed for construction sites, and skilled technicians and engineers will be required for the maintenance of resource deposits. The Russian Arctic is experiencing a labor shortage, exacerbated by Russia's overall negative demographic trends and the war in Ukraine, which has resulted in increased fatalities and a mass exodus of young skilled workers. The Arctic region thus remains central to Moscow's projection of its economic and strategic sovereignty, yet in many aspects—social as well as spatial—it remains one of many among Russia's peripheral spaces.

With Russia's war in Ukraine, Moscow has decoupled from the West and accelerated its interdependency with China, with the balance of power now favoring Beijing. Russia has succeeded in addressing the main pressing issues resulting from sanctions and its transformations into a war economy. Even if its progress toward reaching its development goals has been slowed down, Russia's society and economy have been able to demonstrate resilience. This resilience feature is the backbone of human development amid the harsh conditions of the Arctic, as experienced through centuries of Russia's unique human presence in its northern Siberian landmass. The current war has also led to the world backsliding on key environmental goals, increasing fuel production, and locking in decades of future extractions and emissions, confirming that great power competition undermines efforts to address climate change and continues a twentieth-century vision of the world order unable to address the planet's finite resources.

ARCTIC CHRONOLOGY

1584	Foundation of the city and port of Arkhangelsk
1898	Opening of Arkhangelsk railway station
1912	Founding of Murmansk
1916	Opening of Murmansk railway station
1923	Creation of SLON (Solovetskii Lager Osobogo Naznatcheniia), the first Soviet camp on the White Sea
1924	Creation of the Northern Committee
1926	Decree "On the Proclamation of Lands and Islands Located in the Northern Arctic Ocean as Territory of the USSR"
1928	Launching of the Gulag Northern Siberian Railway to serve the coal deposits of Pechora, the northern Urals, and the middle Ob region
1932	Creation of the Main Directorate of the Northern Sea Route (dissolved in 1964)
1932	The Schmidt expedition: first recognized Arctic passage from west to east in a single navigation season
1934	Joint Resolution of the USSR's Council of People's Commissars and the Central Committee of the All-Union Communist Party of Bolsheviks "On measures for the development of the Northern Sea Route and the northern economy"
1935	Dissolution of the Northern Committee
1935	Start of exploitation of the Norilsk deposit
1935	First commercial passage on the Northern Sea Route
1937	Ratification by the USSR of the Spitsbergen Treaty (aka Svalbard Treaty) with Norway
1937	Launch of the drifting SP-1 Severnyi Polius 1 Base under Soviet polar explorer Ivan Papanin

Arctic Chronology

1937–41	Gulag construction of the Vologda-Pechora (Vorkuta) Railroad Line
1941	First American-British Lend-Lease convoy arrives via the Arctic Route at Arkhangelsk
1942	A branch of the Pacific Lend-Lease Route passes through the Bering Strait to the mouths of the great Siberian rivers
1947–53	Construction by the Gulag of the Chum (Vorkuta)-Labitnangui (Ob) Railroad Line
1949	American "Weather Reconnaissance Squadron" begins daily flights between Alaska and Japan to detect atomic particles
1950	The SP-2 Drifting Base is launched, which will remain in all but uninterrupted operation until 1991
1953	First gas field exploited at Beryozovo (Yamalo-Nenets district)
1954	Opening of the Novaya Zemlya Nuclear Polygon: from 1955 to 1990, between 134 and 224 nuclear explosions take place there
1957	Launch of the *Lenin*, the first Soviet nuclear icebreaker
1958	The USS *Nautilus*, an American nuclear submarine, passes under the ice at the North Pole
1960	List of the Far North and areas equated to the Far North published, subject to the Decrees of the Presidium of the Supreme Soviet of the USSR of February 10, 1960
1960	First oil wells exploited in the Konda Valley (Yamalo-Nenets district)
1961	"Tsar-Bomba," the most powerful bomb ever used, explodes in Novaya Zemlya
1972	The Central Committee of the CPSU decides to build over-the-horizon radars in the Soviet Arctic to prevent US missile attacks
1972	Launch of the *Arktika*, ushering in a new generation of icebreakers
1974	The Soviet Union and the United States signed the Treaty on the Limitation of Underground Nuclear Weapon Tests, also known as the Threshold Test Ban Treaty (TTBT)

1976	The first "Riteg" small-capacity nuclear generators are sent to the Arctic seaboard. More than 1,000 will be commissioned by 1990
1976	Wrangel Island is declared a "USSR reserve," the first in the Soviet Arctic
1977	The *Arktika* becomes the first icebreaker to reach the North Pole
1978	Fishing agreement between Norway and USSR on the so-called Grey Zone in Barents Sea
1982	Adoption of the UN Convention on the Law of the Sea (UNCLOS)
1984	The Pechora over-the-horizon radar goes into service, covering the Arctic
1985	Creation of the Sevmorneftegazgeofizrazvedka trust, tasked with exploring the Barents Sea for oil and gas deposits
1987	Discovery of the Peschanoozerskoye oil field on the island of Kolguyev
1988	Discovery of the Shtokman gas field in the northwestern part of the South Barents Basin
1989	Discovery of the Prirazlomnoye oil field in the Pechora Sea, south of Novaya Zemlya
1989	April: USSR Council of Ministers' State Commission on Arctic Affairs defines the country's Arctic zone
1989	The USSR admits to violating the Anti-Ballistic Missile Treaty by attempting to build over-the-horizon radars covering its territory. Only those in the Pechora (Komi Republic) and Kabala (Azerbaijan) regions are put into service
1989	At the invitation of the government of Finland, first official meeting from the eight Arctic countries in Rovaniemi, Finland
1990	Signing of the Baker-Shevardnadze Treaty, establishing the border between Russia and the United States in the Bering Strait

Arctic Chronology

1990	The Soviet Union proposed a moratorium on nuclear testing that was agreed to by the UK and the United States. The Soviet Union's last nuclear test took place on October 24, 1990
1991	On Approval of the Regulations on the Commission on Arctic and Antarctic Affairs under the Cabinet of Ministers of the USSR and the Personal Composition of This Commission issued (Resolution of the Cabinet of Ministers of the USSR no. 308 of May 29, 1991)
1991	Signature of the Arctic Environmental Protection Strategy in Rovaniemi, Finland
1992	Interdepartmental Commission on Arctic and Antarctic Affairs established, which replaced the RSFSR State Commission on Arctic and Antarctic Affairs
1992	Signing in Rio de Janeiro of the United Nations Framework Convention on Climate Change (UNFCCC)
1996	Program for the development of hydrocarbon reserves on the shelf of Russia's Arctic seas until 2010 approved
1996	Federal Law no. 78 of June 19, 1996, On the Fundamentals of State Regulation of the Socio-Economic Development of the North of the Russian Federation (lapsed in 2005)
1997	Adoption of the Kyoto Protocol on reducing emissions within the framework of the UNFCCC
2000	Sinking of the *Kursk* submarine in the Barents Sea
2000	The Concept of State Support for Economic and Social Development of the Northern Regions (approved by Resolution of the Government of the Russian Federation no. 198 of March 7, 2000)
2001	First Russian submission to UNCLOS
2004	Russia Ratifies Kyoto Protocol on Emissions
2007	Russian titanium flag flown over the North Pole seabed
2008	Dmitri Medvedev's decree, On the Foundations of Russian Arctic Policy until 2020
2009	Establishment of Russian Arctic National Park
2010	Signing of the Russian-Norwegian treaty on the delimitation of their border in the Barents Sea

2012	Creation of the Ministry of Development of the Far East
2012	Duma passes the long-awaited Law on the Northern Sea Route
2013	Strategy for the Development of the Arctic Zone of the Russian Federation and National Security until 2020 (approved by the President of the Russian Federation on February 8)
2013	Reestablishment of the Main Directorate of the Northern Sea Route (placed under Rosatom in 2018)
2013	Creation of Beringia National Park
2014	State program on Social and Economic Development of the Arctic Zone of the Russian Federation (approved by Resolution of the Government of the Russian Federation no. 366 of April 21, 2014)
2014	Presidential Decree no. 296, On Land Territories of the Arctic Zone of the Russian Federation (May 2)
2015	Second Russian submission to UNCLOS, conditionally accepted in 2023
2015	Establishment of State Commission for Arctic Development
2019	State Commission for Arctic Development is attached to the Ministry for the Development of the Far East, which becomes the Ministry for the Development of the Far East and the Arctic
2020	Adoption of the Law on State Support for Entrepreneurial Activity in the Arctic Zone of the Russian Federation
2021	Russia officially takes over chairmanship of the Arctic Council for two years
2022	Full-scale invasion of Ukraine; freeze of the Arctic Council decided by all other members
2023	UNCLOS accepts the main elements of the Russian request to extend its declared continental shelf into the Arctic Ocean
2023	Finland joins NATO
2024	The Permafrost Act comes into force in the Yamalo-Nenets Autonomous District

Arctic Chronology

2024	Sweden joins NATO
2024	Announcement of the creation of a Russo-Chinese commission for the development of the Northern Sea Route
2024	First meeting of the Russo-Indian group on cooperation in the Northern Sea Route

NOTES

Introduction

1 Thane Gustafson, *Klimat: Russia in the Age of Climate Change* (Cambridge, MA: Harvard University Press, 2021).

Chapter 1

1 Parts of this section are updated from Marlene Laruelle, *Russia's Arctic Strategies and the Future of the Far North* (London: Routledge, 2013).
2 More in Alex G. Elferink, "The Law and Politics of the Maritime Boundary Delimitation of the Russian Federation, Part 1," *International Journal of Marine and Coastal Law* 11, no. 4 (1996): 525–61; Alex G. Elferink, "The Law and Politics of the Maritime Boundary Delimitation of the Russian Federation, Part 2," *International Journal of Marine and Coastal Law* 12, no. 1 (1997): 5–35.
3 Alexander A. Sergunin, "Introduction: Quo Vadis? The Arctic between Nationalism and Globalism," *Polar Journal* 11, no. 1 (2021): 1–10.
4 Cornell Overfield, "An Off-the-Shelf Guide to Extended Continental Shelves and the Arctic," Lawfare (blog), April 21, 2021, https://www.lawfareblog.com/shelf-guide-extended-continental-shelves-and-arctic.
5 See the US Notification regarding Russia's submission to the CLCS (March 18, 2002), http://www.un.org/Depts/los/clcs_new/submissions_files/rus01/CLCS_01_2001_LOS__USAtext.pdf. See also Brian Spielman, "An Evaluation of Russia's Impending Claim for Continental Shelf Expansion: Why Rule 5 Will Shelve Russia's Submission," *Emory International Law Review* 23 (2009): 309–49.
6 Tomasz Górski, "A Note on Submarine Ridges and Elevations with Special Reference to the Russian Federation and the Arctic Ridges," *Ocean Development & International Law* 40, no. 1 (2009): 51–60.
7 N. F. Coehlo, "Russia: Revised Submission to the CLCS in the Area of the Gakkel Ridge (Arctic Ocean)," *Demaribus*, December 4, 2023,

Notes

    https://demaribus.net/2023/12/04/russia-revised-submission-to-the-clcs-in-the-area-of-the-gakkel-ridge-arctic-ocean/; Martin Breum, "Russia Gets Approval for the Data Behind Much of Its Arctic Ocean Seabed Claim," *ArcticToday*, February 17, 2023, https://www.arcticto day.com/russia-gets-approval-for-the-data-behind-much-of-its-arctic-ocean-seabed-claim/.

8. Danielle Bochove, "US Claims Huge Chunk of Seabed amid Strategic Push for Resources," *Bloomberg*, December 22, 2023, https://www.bloomberg.com/news/articles/2023-12-22/us-claims-huge-chunk-of-seabed-amid-strategic-push-for-resources; Liam Denning, "US Joins Arctic Race to Grab Resource-Rich Seabed," *Bloomberg*, January 4, 2024, https://www.bloomberg.com/opinion/articles/2024-01-04/us-joins-arctic-race-to-grab-resource-rich-seabed.

9. Congressional Research Service, *Outer Limits of the US Extended Continental Shelf: Background and Issues for Congress* (Washington, DC: Congressional Research Service, February 7, 2024), 10, https://crsreports.congress.gov/product/pdf/R/R47912#:~:text=On%20December%2019%2C%202023%2C%20the,nmi%20(987%2C700%20square%20kilometers.

10. Didier Ortolland, "La guerre en Ukraine déstabilise l'Arctique," *Le Monde diplomatique*, September 2024, https://www.monde-diplomatique.fr/2024/09/ORTOLLAND/67463; Congressional Research Service, *Outer Limits of the U.S. Extended Continental Shelf*; Andrey Todorov, "Russia's Reaction to the US Continental Shelf Announcement: Political Posturing or Setting the Stage for a Big Move?" The Arctic Institute Briefs (blog), April 9, 2024, https://www.thearcticinstitute.org/russias-reaction-us-continental-shelf-announcement-political-posturing-setting-stage-big-move/.

11. Martin Breum, "Russia Gets Approval for the Data behind Much of Its Arctic Ocean Seabed Claim"; Lukas B. Wahden, "Strategic Brief no. 68—2024—Will Russia Denounce the United Nations Convention on the Law of the Seas [sic] (UNCLOS)?" Institut de Recherche Stratégique de l'École Militaire, IRSEM Strategic Brief 68, May 14, 2024, https://www.irsem.fr/strategic-brief-no-68-2024.html.

12. "Agreement between the United States of America and the Union of Soviet Socialist Republics on the Maritime Boundary," June 1, 1990, United Nations, http://www.un.org/Depts/los/LEGISLATIONANDTREATIES/PDFFILES/TREATIES/USA-RUS1990MB.PDF.

13. Spielman, "An Evaluation of Russia's Impending Claim for Continental Shelf Expansion," 339.

14. Arild Moe, "The Russian Barents Sea: Openings for Norway?," in *High North: High Stakes*, ed. Rosa Gottemoeller and R. Tamnes (Bergen: Fagbokforlaget, 2008), 75–85.

15 Kristoffer Stabrun, "The Grey Zone Agreement of 1978 Fishery Concerns, Security Challenges and Territorial Interests," *FNI Report* 13 (2009), www.files.ethz.ch/isn/112916/FNI-R1309.pdf.

16 Øystein Jensen and Svein Vigeland Rottem, "The Politics of Security and International Law in Norway's Arctic Waters," *Polar Record* 46, no. 236 (2010): 75–83; Andreas Østhagen, "High North, Low Politics—Maritime Cooperation with Russia in the Arctic," *Arctic Review on Law and Politics* 7, no. 1 (2016): 83–100.

17 "Treaty between the Kingdom of Norway and the Russian Federation Concerning Maritime Delimitation and Cooperation in the Barents Sea and the Arctic Ocean," English version, September 15, 2010, http://www.kremlin.ru/ref_notes/707, http://www.regjeringen.no/upload/SMK/Vedlegg/2010/avtale_engelsk.pdf.

18 Tore Henriksen and Geir Ulfstein, "Maritime Delimitation in the Arctic: The Barents Sea Treaty," *Ocean Development & International Law* 42, nos. 1–2 (2011): 1–21.

19 Arild Moe, D. Fjærtoft, and I. Øverland, "Space and Timing: Why Was the Barents Sea Delimitation Dispute Resolved in 2010?" *Polar Geography* 34, no. 3 (2011): 145–62.

20 D. H. Anderson, "The Status under International Law of the Maritime Areas around Svalbard," *Ocean Development & International Law* 40, no. 4 (2009): 373–84.

21 A. N. Vylegzhanin and V. L. Zilanov, *Spitsbergen: Legal Regime of Adjacent Marine Areas* (Portland, OR: Eleven International, 2007), 57.

22 Kristian Åtland and Torbjørn Pedersen, "The Svalbard Archipelago in Russian Security Policy: Overcoming the Legacy of Fear—or Reproducing It?" *European Security* 17, no. 2 (2008): 227–51.

23 Mathieu Landriault, "Russia and the Immediate Future of Arctic Geopolitics," Centre for International Policy Studies, University of Ottawa, July 23, 2024, https://www.cips-cepi.ca/2024/07/23/russia-and-the-immediate-future-of-arctic-geopolitics/.

24 Mia M. Bennett et al., "The Opening of the Transpolar Sea Route: Logistical, Geopolitical, Environmental, and Socioeconomic Impacts," *Marine Policy* 121 (2020): 1–15, https://par.nsf.gov/servlets/purl/10191982.

25 Laurent Fedi, Laurent Etienne, and Olivier Faury, "Mapping and Analysis of Maritime Accidents in the Russian Arctic through the Lens of the Polar Code and POLARIS System," *Marine Policy* 118 (August 2020): 1–28, https://hal.science/hal-02885952/document.

26 W. E. Butler, *The Northeast Arctic Passage* (Alphen aan den Rijn: Sijthoff & Noordhoof International, 1978).

Notes

27 "Zakon 'O vnesenii izmenenii v otdel'nye zakonodatel'nye akty RF v chasti gosudarstvennogo regulirovaniia torgovogo moreplavaniia v akvatorii Severnogo morskogo puti," July 28, 2012, *Rossiiskaia gazeta*, July 30, 2012, http://www.rg.ru/2012/07/30/more-dok.html.

28 On the complex legal debate on extraterritoriality of EU and US shipping laws and how Russia does or does not apply them, see Igor V. Stepanov and Peter Ørebech, *Legal Implications for the Russian Northern Sea Route and Westward in the Barents Sea* (Oslo: Fridtjof Nansens Institutt, 2005).

29 Leonid Timtchenko, "The Russian Arctic Sectoral Concept: Past and Present," *Arctic* 50, no. 1 (1997): 29–35.

30 Government of the Russian Federation, "O gosudarstvennoi podderzhke predprinimatel'skoi deiatel'nosti v Arkticheskoi zone Rossiiskoi Federatsii," July 7, 2020, http://publication.pravo.gov.ru/Document/View/0001202007130047?ysclid=lw2afr6tmt991261739.

31 Jean Radvanyi, *La nouvelle Russie*, 4th edition (Paris: Armand Colin, 2010), chap. 15–16.

32 Yuri Slezkine, *Arctic Mirrors: Russia and the Small Peoples of the North* (Ithaca, NY: Cornell University Press, 1994).

33 Asif Siddiqi, "Atomized Urbanism: Secrecy and Security from the Gulag to the Soviet Closed Cities," *Urban History* 49, no. 1 (2022): 190–210.

34 Varlam Shalamov, *Kolyma Tales* (New York: Penguin, [1954–73] 1995); Alexander Solzhenitsyn, *One Day in the Life of Ivan Denisovich* (New York: Signet, [1962] 2008).

35 Dmitry Arzyutov, "The Making of the *Homo Polaris*: Human Acclimatization to the Arctic Environment and Soviet Ideologies in Northern Medical Institutions," *Settler Colonial Studies* 14, no. 2 (2023): 180–203.

36 Nikolai I. Shiklomanov, "All Fall Down? Urban Infrastructure and Permafrost in the Russian Arctic," *Russian Analytical Digest* 261 (2020): 8, https://css.ethz.ch/content/dam/ethz/special-interest/gess/cis/center-for-securities-studies/pdfs/RAD261.pdf.

37 See Marlene Laruelle's and Nikolai Shiklomanov's special issue on Norilsk in *Polar Geography* 40, no. 4 (2017): 251–6. https://www.tandfonline.com/toc/tpog20/40/4. See also Taline Ter Minassian, *Norilsk: L'architecture au Goulag* (Paris: B2 Publisher Paris, 2018).

38 Shiklomanov, "All Fall Down?" 8.

39 National Aeronautics and Space Administration, "Arctic and Antarctic Sea Ice Approached Historic Lows," NASA Earth Observatory,

September 11, 2024, https://earthobservatory.nasa.gov/images/153457/arctic-and-antarctic-sea-ice-approached-historic-lows#:~:text=Arctic%20sea%20ice%20retreated%20to,Ice%20Data%20Center%20(NSIDC.

40 Anne Gädeke, Moritz Langer, Julia Boike, Eleanor J Burke, Jinfeng Chang, Melissa Head, Christopher P O Reyer, Sibyll Schaphoff, Wim Thiery, and Kirsten Thonicke, "Climate Change Reduces Winter Overland Travel across the Pan-Arctic even under Low-End Global Warming Scenarios," *Environmental Research Letters* 16, no. 2 (2021) 14 pages, https://iopscience.iop.org/article/10.1088/1748-9326/abdcf2.

41 Ellen Gray, "Unexpected Future Boost of Methane Possible from Arctic Permafrost," Climate NASA, August 20, 2018, https://science.nasa.gov/earth/climate-change/unexpected-future-boost-of-methane-possible-from-arctic-permafrost/.

42 National Oceanographic and Atmospheric Administration, Arctic Report Card: Update for 2024, NOAA website, https://arctic.noaa.gov/report-card/.

43 Risto K. Heikkinen, Mista Luoto, Raimo Virkkala, and K. Rainio, "Effects of Habitat Cover, Landscape Structure and Spatial Variables on the Abundance of Birds in an Agricultural-Forest Mosaic," *Journal of Applied Ecology* 41, no. 5 (2004): 824–35.

44 Ruonan Wu, Gareth Trubl, Neslihan Taş, and Janet K. Jansson, "Permafrost as a Potential Pathogen Reservoir," *One Earth* 5, no. 4 (2022): 351–60.

45 Tat'iana Uskova, "Treshchiny, vspuchivanie i 'vorota i ad': chem grozit Rossii taianie vechnoi merzloty," *MBX*, October 23, 2019, https://mbk-news.appspot.com/suzhet/chem-grozit-rossii-tayanie-vechnoj-merzloty/.

46 Marlene Laruelle, Igor Esau, Martin Miles, Victoria Miles, Anna N. Kurchatova, Sergej A. Petrov, Andrey Soromotin, Mikhail Varentsov and Pavel Konstantinov, "Arctic Cities as an Anthropogenic Object: A Preliminary Approach through Urban Heat Islands," *Polar Journal* 9, no. 2 (2019): 402–23.

47 Dmitri A. Streletskiy et al., "Assessment of Climate Change Impacts on Buildings, Structures and Infrastructure in the Russian Regions on Permafrost," *Environmental Research Letters* 14, no. 2 (2019), https://iopscience.iop.org/article/10.1088/1748-9326/aaf5e6.

48 Oleg Anisimov and Svetlana Reneva, "Permafrost and Changing Climate: The Russian Perspective," *Ambio* 35, no. 4 (2006): 169–75.

49 Elena Wilson Rowe, *Russian Climate Politics* (New York: Palgrave Macmillan, 2014).

Notes

50 The Economist, "Why Russia Is Ambivalent about Global Warming?" *The Economist*, September 19, 2019, https://www.economist.com/europe/2019/09/19/why-russia-is-ambivalent-about-global-warming?utm_medium=cpc.adword.pd&utm_source=google&ppccampaignID=18151738051&ppcadID=&utm_campaign=a.22brand_pmax&utm_content=conversion.direct-response.anonymous&gad_source=1&gclid=Cj0KCQiAsaS7BhDPARIsAAX5cSD57Xm8FEgPVZWUMcwxKVVTOxSXOOGJmaLyhQjnS67f7onjLEke9hQaAjZBEALw_wcB&gclsrc=aw.ds.

51 Quoted from *The Moscow Times*, "Skepticism to Acceptance: How Putin's Views on Climate Change Evolved Over the Years," *The Moscow Times*, July 1, 2021, https://www.themoscowtimes.com/2021/07/01/skepticism-to-acceptance-how-putins-views-on-climate-change-evolved-over-the-years-a74391. See also Barbara Buchner and Silvia Dall'Olio, "Russia and the Kyoto Protocol: The Long Road to Ratification," *ENP Environment Networks Papers* 12 (2005): 349–82.

52 Steven Lee Myers, "Putin Ratifies Kyoto Protocol on Emissions," *New York Times*, November 6, 2004, https://www.nytimes.com/2004/11/06/world/europe/putin-ratifies-kyoto-protocol-on-emissions.html.

53 Prezident Rossiiskoi Federatsii, "Plenarnoe zasedanie Mezhdunarodnogo arkticheskogo foruma," *Kremlin.ru*, April 9, 2019, http://www.kremlin.ru/events/president/transcripts/statements%20/60250

54 Igor Dmitrov, "Putin rasskazal ob ugroze rossiiskim gorodam v Arktike iz-za potepleniia klimata," Lenta.ru, June 4, 2021, https://lenta.ru/news/2021/06/04/ugroza/?ysclid=lw53stjo92407195700.

55 Ellie Martus, "Policymaking and Policy Framing: Russian Environmental Politics under Putin," *Europe-Asia Studies* 73, no. 5 (2021): 869–89.

56 RBC, "Vygody Rossii ot global'nogo potepleniia otsenili bolee chen v 1 trln rublei," *RBC*, August 30, 2024, rbc.ru/economics/30/08/2024/66d0576c9a7947eecd47122f.

57 Alexander Sergunin and Valerii Konyshev, "Russia's Arctic Strategy," in *Russia Strategy, Policy and Administration*, ed. Irvin Studin (London: Palgrave Macmillan, 2018), 135–44.

58 Moscow staff, "Putin Orders State of Emergency after Huge Fuel Spill Inside Arctic Circle," *Guardian*, June 3, 2020, https://www.theguardian.com/environment/2020/jun/03/vladimir-putin-orders-state-of-emergency-huge-fuel-spill-siberia-power-plant-kerch.

59 Igor Esau and Victoria Miles, "Warmer Urban Climates for Development of Green Spaces in Northern Siberian Cities," *Geography, Environment, Sustainability* 9, no. 4 (2016): 48–62.

60 Nadja M. Tchebakova, Elena Parfenova, and Amber J. Soja, "The Effects of Climate, Permafrost and Fire on Vegetation Change in Siberia in a Changing World," *Environmental Research Letters* 4, no. 4 (2009): 1–9.

61 Rachael Treharne et al., "Arctic Browning: Impacts of Extreme Climatic Events on Heathland Ecosystem CO_2 Fluxes," *Global Change Biology* 25, no. 2 (February 2019): 489–503.

62 GeoInfo, "V Iamale vstupil v zakonnuiu silu zakon ob obiazatel'nom geokriologicheskom monitoring i okhrane merzloty," GeoInfo.ru, January 9, 2024, https://geoinfo.ru/product/sluzhba-novostej-geoinfo/Yamale-vstupil-v-zakonnuyu-silu-zakon-ob-obyazatelnom-geokriologicheskom-monitoringe-i-ohrane-merzloty-51192.shtml?ysclid=lweqlhqtlt757421358.

63 Terence Armstrong, *The Northern Sea Route: Soviet Exploitation of the North East Passage* (Cambridge: Cambridge University Press, 2011).

64 Listed in William Dunlap, *Transit Passage in the Russian Arctic Straights* (Durham: International Boundaries Research Unit, University of Durham, 2002), 26–34.

65 "Lukoil, Iukos, i TNK postroiat port v Arktike za 1,5 milliard dollarov," Neftegaz.ru, December 9, 2002, https://neftegaz.ru/news/companies/317954-neftyanoy-port-v-arktike/?ysclid=lwj8idulrz747680785; Sabrina Tavernise, "Russia Plans Oil Pipeline to Arctic Port," *New York Times*, November 28, 2002, https://www.nytimes.com/2002/11/28/business/russia-plans-oil-pipeline-to-arctic-port.html.

66 "Ob"em gruzoperevozok po Sevmorputi dostig rekordnykh 37.3 mln ton na god," *Vzgliad*. December 28, 2024, https://vz.ru/news/2024/12/28/1306184.html and German Kostrinskii, "Perevozki po Severnomu puti v 2024 g. otkloniatsia ot plana vdvoe," *RBC*, March 29, 2024, https://www.rbc.ru/business/29/03/2024/6606cd9a9a7947b91495e22e?ysclid=m5aultjxiw909414098.

67 Olga Solovieva, "K 2030 godu vlasti obeshchaiut spustit' na vodu novyi arkticheskii flot," *Nezavisimaia gazeta*, May 20, 2024, https://www.ng.ru/economics/2024-05-20/4_9011_fleet.html.

68 German Kostrinskii, "Perevozki po Severnomu puti v 2024 godu otkloniatsia ot plana vdvoe."

69 GoArctic, "Skol'ko i kakikh gruzov, otkuda, kuda i kakimi sudami perevezli po Severnomu morskomu puti, i chto mozhno ozhidat' v dal'neishem," GoArctic.ru, December 23, 2024, https://goarctic.ru/work/skolko-i-kakikh-gruzov-otkuda-kuda-i-kakimi-sudami-perevezli-po-severnomu-morskomu-puti-i-chego-mozh/?ysclid=m5ax7ot85j524873104.

70 Tatiana Sorokina and William G. Phalen, "Legal Problems of the Northern Sea Route Exploitation: Brief Analysis of the Legislation of

Notes

the Russian Federation," in *International Marine Economy: Law and Policy*, ed. Myron H. Nordquist, John Norton Moore, and Ronán Long (Leiden: Brill, 2017), 99–120; Brandon M. Boylan, "Increased Maritime Traffic in the Arctic: Implications for Governance of Arctic Sea Routes," *Marine Policy* 131 (2021): no pagination, https://www.sciencedirect.com/science/article/abs/pii/S0308597X21001779.

71 Prezident Rossiiskoi Federatsii, "Podpisan zakon o nadelenii 'Rosatoma' riadom polnomychii v oblasti razvitiia Severnogo morskogo puti," Kremlin.ru, December 28, 2018, http://www.kremlin.ru/acts/news/59539.

72 Atle Staalesen, "Russian Legislators Ban Foreign Shipments of Oil, Natural Gas and Coal Along Northern Sea Route," Barents Observer (news site), December 26, 2017, https://www.thebarentsobserver.com/arctic/russian-legislators-ban-foreign-shipments-of-oil-natural-gas-and-coal-along-northern-sea-route/115574.

73 Maritime Executive, "Russia Tightens Control over Northern Sea Route," MaritimeExecutive.com, March 8, 2019, https://maritime-executive.com/article/russia-tightens-control-over-northern-sea-route; Warsaw Institute, "Russia Imposes Foreign Sailing Restrictions on Northern Sea Route," WarsawInstitute.org, March 8, 2019, https://warsawinstitute.org/russia-imposes-foreign-sailing-restrictions-northern-sea-route/; see also Alexey Kozachenko, Bogdan Stepovoi, Elnar Bainazarov, "Kholodnaia volna: inostrantsam sozdali pravila prokhoda Sevmorputi," *Izvestiia*, March 6, 2019, https://iz.ru/852943/aleksei-kozachenko-bogdan-stepovoi-elnar-bainazarov/kholodnaia-volna-inostrantcam-sozdali-pravila-prokhoda-sevmorputi.

74 "Russia to Spend 1.8 Trillion Rubles on Northern Sea Route Development by 2035," Arctic.ru, August 4, 2022, https://arctic.ru/economics/20220804/1003501.html.

75 Interfax, "Rosatom i NOVATEK v 2024 g. zapustiat kruglogodichnuiu navigatsiiu na vostoke po Sevmorputi," Interfax, May 17, 2023, https://www.interfax.ru/russia/901869.

76 Arctic Russia, "Levodyi navigator za tri goda. 'Novyi kosmos' i chastnye resheniia dlia SMR," ArcticRussia.ru, March 27, 2024, https://arctic-russia.ru/article/ledovyy-navigator-za-tri-goda-novyy-kosmos-i-chastnye-resheniya-dlya-smp//.

77 Dzen, "Severnyi shirotnyi sdaet zadnym khodom," dzen.ru, July 24, 2023, https://dzen.ru/a/ZL48EpaHugWO2rno?ysclid=m5nysz0c50630771004.

Notes

Chapter 2

1 See examples of these multifaceted cooperation in Alexander A. Sergunin, ed., *Handbook of Research on International Collaboration, Economic Development, and Sustainability in the Arctic* (Chicago: IGI Global, 2019).

2 Kristian Åtland, "Mikhail Gorbachev, the Murmansk Initiative, and the Desecuritization of Interstate Relations in the Arctic," *Cooperation and Conflict* 43, no. 3 (2008): 289–311.

3 The countries in question were: the United States, Russia, Finland, Norway, Japan, China, Canada, and South Korea. See Vladimir Vasiliev, Boris Krasnopolsky, and Alexander Piliasov, *Rozhdennyi ob"edeniat' (k 30 letiu Severnogo Forum)* (Moscow-Smolensk: Universum, 2023).

4 Douglas C. Nord, *The Arctic Council: Governance within the Far North* (London: Routledge, 2016).

5 Piotr Graczyk and Svein Vigeland Rottem, "The Arctic Council: Soft Actions, Hard Effects?" in *The Routledge Handbook of Arctic Security*, ed. Gunhild Hoogensen Gjørv, Marc Lanteigne, and Horatio Sam-Aggrey (London: Routledge, 2020): 221–34.

6 Alexander Sergunin, "Thinking about Forthcoming Russian Arctic Council Chairmanship: Challenges and Opportunities," *Polar Science* 29 (2021): 1–9.

7 Torbjørn Pedersen and Beate Steinveg, "Russia's Clashing Ambitions: Arctic Status Quo and World-Order Revision," *Politics and Governance* 12 (2024): no pagination.

8 Torbjørn Pedersen, "Debates over the Role of the Arctic Council," *Ocean Development & International Law* 43, no. 2 (2012): 146–156; Kathryn C. Lavelle, "Regime, Climate, and Region in Transition: Russian Participation in the Arctic Council," *Problems of Post-Communism* 69, nos. 4–5 (2022): 345–57.

9 Oran R. Young, "Arctic Politics in an Era of Global Change," *Brown Journal of World Affairs* 19, no. 1 (2012): 165–78.

10 Heather Exner-Pirot and Robert W. Murray, "Regional Order in the Arctic: Negotiated Exceptionalism," *Politik* 20, no. 3 (2017): 47–64; Pavel Devyatkin, "Arctic Exceptionalism: A Narrative of Cooperation and Conflict from Gorbachev to Medvedev and Putin," *Polar Journal* 13, no. 2 (2023): 336–57.

11 Valery Konyshev and Alexander Sergunin, "Is Russia a Revisionist Military Power in the Arctic?" *Defense & Security Analysis* 30, no. 4 (2014): 323–35.

Notes

12 Lassi Heininen, "Arctic Geopolitics from Classical to Critical Approach—Importance of Immaterial Factors," *Geography, Environment, Sustainability* 11, no. 1 (2018): 171–86.

13 Nele Matz-Lück, "Planting the Flag in Arctic Waters: Russia's Claim to the North Pole," *Göttingen Journal of International Law* 1, no. 2 (2009): 235–55.

14 Tom Parfitt, "Russia Plants Flag on North Pole Seabed," *Guardian*, August 2, 2007, http://www.guardian.co.uk/world/2007/aug/02/russia.arctic.

15 Artur Chilingarov, "Arktika—nash rodnoi krai," Regnum.ru, July 7, 2007, http://www.regnum.ru/news/867158.html.

16 "SShA i Rossiia razdeliaiut Arktiku," Pogranichnik.ru, January 14, 2009, http://forum.pogranichnik.ru/index.php?showtopic=10737.

17 Alicja Curanović, "The Keeper of the Imperial Body: The Russian Geographical Society as an Entrepreneur of Imperial Nationalism," *Nationalities Papers* 53, no. 2 (2024): 1–19.

18 Quoted From Maria Antonova, "State Lays Claim to Academic Society," *The Moscow Times*, November 19, 2009, https://www.themoscowtimes.com/archive/state-lays-claim-to-academic-society. See also Jean Radvanyi, "When Putin Turned Geographer," *Hérodote*, no. 166 (2017): 113–32. There is no volume number.

19 Østhagen, 2021, M. Michael Byers (2017). "Crisis and International Cooperation: An Arctic Case Study," *International Relations* 31, no. 4 (2017), 1–28; Andreas Østhagen, "Norway's Arctic Policy: Still High North, Low Tension?" *Polar Journal* 11, no. 1 (2021): 75–94.

20 Pavel Baev, "Russia's Ambivalent Status-Quo/Revisionist Policies in the Arctic," *Arctic Review on Law and Politics* 9 (2018): 408–24.

21 TASS, "Shoigu: Arktika stala tsentrom interesov riada gosudarstv, chto mozhet privesti k konfliktam" [Shoigu: The Arctic has become a center of interest for several states, which could lead to conflict], Russian News Agency TASS, August 31, 2018, https://tass.ru/armiya-i-opk/5509944.

22 Elana Wilson Rowe, "Analyzing Frenemies: An Arctic Repertoire of Cooperation and Rivalry," *Political Geography* 76 (2020): 10, https://www.sciencedirect.com/science/article/pii/S0962629818305158.

23 Arctic Council, "Russian Chairmanship 2021–2023," https://arctic-council.org/about/previouschairmanships/russian-chairmanship-2; Alexander Sergunin, "Thinking about Russian Arctic Council Chairmanship: Challenges and Opportunities," *Polar Science* 29 (2021): 9.

24 Barry Scott Zellen, "The Arctic Council Pause: The Importance of Indigenous Participation and the Ottawa Declaration," *Arctic Circle*,

2022, https://www.arcticcircle.org/journal/the-importance-of-indigenousparticipation-and-the-ottawa-declaration.

25 Arctic Council, Agreement on Cooperation on Marine Oil Pollution Preparedness and Response in the Arctic, 2013, https://oaarchive.arctic-council.org/items/ee4c9907-7270-41f6-b681-f797fc81659f.

26 Jari Tanner, "Norway Is Mulling Building a Fence on Its Border with Russia, Following Finland's Example," AP News, September 29, 2024, https://apnews.com/article/norway-russia-border-fence-finland-migrants-arctic-dad4878a24fa550dd9eac9d1a6532274.

27 Trine Jonassen, "Russia Will Stay in the Arctic Council as Long as It Serves Our Interests," High North News, May 11, 2023, https://www.highnorthnews.com/en/russia-will-stay-arctic-council-long-it-serves-our-interests.

28 Nikita Lipunov and Pavel Devyatkin, "The Arctic in the 2023 Russian Foreign Policy Concept," The Arctic Institute, May 30, 2023, https://www.thearcticinstitute.org/arctic-2023-russian-foreign-policy-concept/.

29 President of the Russian Federation, "Changes to Basic Principles of State Policy in the Arctic until 2035," Kremlin.ru, February 21, 2023, http://en.kremlin.ru/acts/news/70570.

30 RIA Novosti, "V MID nazvali uslovie dlia vykhoda iz Arkticheskogo soveta," RIA Novosti (news agency), February 6, 2024, https://ria.ru/20240206/arkticheskiy_sovet-1925595556.html.

31 Tatiana Yu. Sorokina, "Pollution and Monitoring in the Arctic," in *Global Arctic. An Introduction to the Multifaceted Dynamics of the Arctic*, ed. Matthias Finger and Gunnar Rekvig (Berlin: Springer Nature, 2022), 229–53.

32 Pavel K. Baev, "Threat Assessments and Strategic Objectives in Russia's Arctic Policy," *Journal of Slavic Military Studies* 32, no. 1 (2019): 25–40.

33 Emelie Moregård, "Nordic Response 2024—NATO Returns to the North in Large Scale," *Northern European and Transatlantic Security (NOTS)*, FOI Memo, 6 pages, April 2024, https://www.foi.se/rest-api/report/FOI%20Memo%208504.

34 Kremlin website, "Ukaz Prezidenta Rossiiskoi Federatsii ot 21.12.2020 no. 803 'O Severnom Flote'," Kremlin.ru, December 21, 2020, http://www.kremlin.ru/acts/bank/46217.

35 Igor Delanoe, "La marine russe et l'Arctique: Nouvelle réalité, anciens enjeux," Le Rubicon (news site), August 18, 2023, https://lerubicon.org/la-marine-russe-et-larctique-nouvelle-realite-anciens-enjeux/.

36 Pavel K. Baev, "Is Russia Really Cutting Its Military Spending?" Jamestown Foundation, Eurasia Daily Monitor, May 6, 2019,

Notes

https://jamestown.org/program/is-russia-really-cutting-its-military-spending/.

37 Pavel K. Baev, "Another Russian Sea Tragedy: Unlearned Lessons Obscured by Secrecy," Jamestown Foundation, Eurasia Daily Monitor, July 8, 2019, https://jamestown.org;https://jamestown.org/program/another-russian-sea-tragedy-unlearned-lessons-obscured-by-secrecy/; Leonid Bershidsky, "Russia Has Failed Another Nuclear Test," *Bloomberg*, August 12, 2019, https://www.bloomberg.com;https://www.bloomberg.com/view/articles/2019-08-12/russia-s-missile-explosion-means-it-failed-two-tests?srnd=opinion; Emma Beswick, "Five Confirmed Dead in an Explosion at a Military Testing Site in Northern Russia," Euronews.com, August 10, 2019, https://www.euronews.com/2019/08/08/two-dead-in-explosion-at-military-testing-site-in-northern-russia-defence-ministry.

38 High North News, "Russia's Forces in the High North: Weakened by the War, Yet Still a Multidomain Threat," High North News, January 12, 2024, https://www.highnorthnews.com/en/russias-forces-high-north-weakened-war-yet-still-multidomain-threat.

39 Malte Humpert, "Ukraine War Taking Toll on Arctic Material and Personnel," High North News, March 13, 2024, https://www.highnorthnews.com/en/ukraine-war-taking-toll-arctic-material-and-personnel.

40 Colin Wall and Njord Wegge, "The Russian Arctic Threat Consequences of the Ukraine War," Center for Strategic & International Studies, CSIS Briefs, January 2023, https://www.jstor.org/stable/pdf/resrep47094.pdf?acceptTC=true&coverpage=false&addFooter=false.

41 Isabel van Brugen, "Russia's Arctic Brigade 'Decimated' in Dnieper Estuary Clashes: Report," *Newsweek*, September 5, 2024, https://www.newsweek.com/russia-arctic-brigade-dnieper-river-clashes-1948686.

42 Mathieu Boulègue, "Russia's Military Posture in the Arctic: Managing Hard Power in a 'Low Tension' Environment," Chatham House Research Paper, October 15, 2024, https://www.chathamhouse.org/2019/06/russias-military-posture-arctic/3-military-infrastructure-and-logistics-russian-arctic.

43 Thomas Nilsen, "Shoigu Vows More Troops Near Nordic Countries," Barents Observer (news site), December 21, 2022, https://thebarentsobserver.com/en/security/2022/12/shoigu-vows-more-troops-near-nordic-countries.

44 Ministry of Foreign Affairs of the Russian Federation, *The Concept of the Foreign Policy of the Russian Federation*, March 31, 2023, https://mid.ru/en/foreign_policy/fundamental_documents/1860586/.

Notes

45 Florian Vidal, "Russia in the Arctic: The End of Illusions and the Emergence of Strategic Realignments," Institut Français des Relations Internationales (IFRI), July 31, 2024, https://www.ifri.org/en/papers/russia-arctic-end-illusions-and-emergence-strategic-realignments.

46 Njord Wegge, "The Strategic Role of Land Power on NATO's Northern Flank," *Arctic Review on Law and Politics* 1 (2022): 94–113, https://www.jstor.org/stable/48710660; Minna Ålander and William Alberque, "NATO's Nordic: Contingency Planning and Learning Lessons," War on the Rocks, December 8, 2022, https://warontherocks.com/2022/12/natos-nordic-enlargement-contingency-planning-and-learning-lessons/

47 Sean Carberry, "US Army Comes Ashore in NATO's High North," *National Defense Magazine*, July 3, 2024, https://www.nationaldefensemagazine.org/articles/2024/7/3/us-army-comes-ashore-in-natos-high-north; High North News, "Regaining Arctic Expertise: US Troops in Alaska Making Strides to Become the Army's Arctic Force," High North News, February 29, 2024, https://www.highnorthnews.com/en/regaining-arctic-expertise-us-troops-alaska-making-strides-become-armys-arctic-force.

48 Asti Evardsen, trans. Brigitte Annie Molid Martinussen, "Pentagon's Upcoming New Arctic Strategy: We Talk Norway, Finland and Sweden a Lot," High North News, May 2, 2024, https://www.highnorthnews.com/en/pentagons-upcoming-new-arctic-strategy-we-talk-norway-finland-and-sweden-lot.

49 Asti Evardsen, trans. Brigitte Annie Molid Martinussen, "All Clear for Nuclear Testing at Novaya Zemlya, Says Russian Head of Test Site," High North News, September 21, 2024, https://www.highnorthnews.com/en/all-clear-nuclear-testing-novaya-zemlya-says-russian-head-test-site.

50 Biznes Vedomosti, "Maluiu atomnuiu stantsiu 'Rosatom' v Iakutii otsenili v 75 mlrd rublei," *Biznes Vedomosti*, September 20, 2024, https://www.vedomosti.ru/business/articles/2024/09/20/1063407-maluyu-atomnuyu-stantsiyu-rosatoma-v-yakutii-otsenili-v-75-mlrd.

51 "World Icebreakers Overview," Chuck Hill's CG Blog, March 1, 2024, https://chuckhillscgblog.net/2024/03/31/world-icebreakers-overview-aker-arctic/.

52 Prezident Rossii, "Tseremoniia spuska na vody atomnogo ledokola 'Chukotka,'" Kremlin.ru, November 6, 2024, http://www.kremlin.ru/events/president/news/75504.

53 Nikolai Petrov, "What Russia's Planned Super Icebreaker Tells Us About Its New Strategic Goals," *Russia.Post*, March 11, 2024, https://russiapost.info/politics/icebreaker

Notes

54 Heiner Kubny, "Tanker without Ice Class on the Northern Sea Route Provokes Criticism," *Polar Journal*, October 12, 2023, https://polarjournal.ch/en/2023/10/12/tanker-without-ice-class-on-the-northern-sea-route-provokes-criticism/.

55 Olga V. Alexeeva and Frédéric Lasserre, "China and the concept of the Third Pole," *Politique etrangère* no. 2 (2022): 177–89. No volume number.

56 M. Taylor Fravel, Kathryn Lavelle, and Liselotte Odgaard, "China Engages the Arctic: A Great Power in a Regime Complex," *Asian Security* 18, no. 2 (2021): 138–58.

57 Marc Lanteigne, "Considering the Arctic as a Security Region: The Roles of China and Russia," in *The Routledge Handbook of Arctic Security*, ed. Gunhild Hoogensen Gjørv, Marc Lanteigne, and Horatio Sam-Aggrey, (London: Routledge, 2017), 311–23; Andrew Chater, "China as Arctic Council Observer: Compliance and Compatibility," North American and Arctic Defense and Security Network, May 3, 2021, https://www.naadsn.ca/wp-content/uploads/2021/05/China%E2%80%99s-Strategic-Objectives-in-the-Arctic-Region-AC-Final.pdf.

58 Yana V. Leksyutina, "China's Participation in Energy Cooperation with Russia in the Arctic," in *Energy of the Russian Arctic*, ed. Valery I. Salygin (Berlin: Springer, 2022), 125–40.

59 Liselotte Odgaard, "Home versus Abroad: China's Differing Sovereignty Concepts in the South China Sea and the Arctic," *Cambridge Review of International Affairs* 37, no. 1 (2024): 60–70; Jingchao Peng and Njord Wegge, "China and the Law of the Sea: Implications for Arctic Governance," *Polar Journal* 2 (2014): 11–15.

60 Elizabeth Wishnik, "The China- Russia 'No Limits' Partnership Is Still Going Strong," Center for Naval Analyses, October 12, 2022, https://www.cna.org/our-media/indepth/2022/10/the-china-russia-no-limits-partnership-is-still-going-strong.

61 Bloomberg News, "Nornickel moves closer to deal for new copper plant in China," *Mining.com*, November 12, 2024, https://www.mining.com/web/norilsk-nickel-moves-closer-to-deal-for-new-copper-plant-in-china/.

62 Robert Fife and Steven Chase, "China Gains Major Arctic Foothold as Russia Turns to Beijing More, Report Finds," *Globe and Mail*, February 7, 2024, https://www.theglobeandmail.com/politics/article-china-arctic-russia-war/.

63 Atle Staalesen, "Russian Arctic Regions Strengthen Bonds with Beijing," Barents Observer, September 20, 2023, https://www.thebarentsobserver.com/arctic/russian-arctic-regions-strengthen-bonds-with-beijing/118562; Zhang Xiaomin, "Dalian to Strengthen Ties with Russian City," *China Daily Global*, May 20, 2024, http://epaper.chinadaily.com.cn/a/202405/20/WS664a8762a310df4030f51a03.html; Ruslan Kuchmov, "Murmanskaia oblast' kak tsentr pritiazheniia kitaiskogo turizma v Rossiiskoi Federatsii," *Upravlencheskoe konsul'tirovanie*, no. 2 (2022): 165–73. No volume number.

64 TASS, "Rossiia i KNP sozdaiut komissiu po razvitiiu SMP," Russian News Agency TASS, May 16, 2024, https://tass.ru/ekonomika/20819733.

65 Lucy Hine, "New Record Shaping Up for 2024 Northern Sea Route Transits," TradeWindsNews.com, October 4, 2024, https://www.tradewindsnews.com/gas/new-record-shaping-up-for-2024-northern-sea-route-transits/2-1-1719840.

66 Noah Bovenizer, "Russia and China to Cooperate on Arctic Shipping Route," ShipTechnology.com, June 7, 2024, https://www.ship-technology.com/news/russia-china-cooperate-arctic-shipping-route/?cf-view.

67 Abbie Tingstad, Stephanie Pezard, and Yulia Shokh, "China-Russia Relations in the Arctic: What Are the Northern Limits of Their Partnership?" RAND Corporation, RAND Expert Insights, November 7, 2024, https://www.rand.org/pubs/perspectives/PEA2823-1.html.

68 Lukas B. Wahden, "Big Words, Small Deeds: Russia and China in the Arctic," Institut de Recherche Stratégique de l'École Militaire (France), IRSEM Research Paper no. 141, February 28, 2024, https://www.irsem.fr/media/5-publications/nr-irsem-141-wahden.pdf.

69 Yury Akimov, "Arctic Paradiplomacy of the Republic of Sakha (Yakutia): The Impact of Federalism, Nationalism, and Identity," in *Mapping Arctic Paradiplomacy: Limits and Opportunities for Sub-National Actors in Arctic Governance*, ed. Mathieu Landriault, Jean-François Payette, and Stéphane Roussel (London: Routledge, 2022), 77–98.

70 Malte Humpert, "Lacking Own Satellite Coverage Russia Is Looking to China for Northern Sea Route Data," High North News, March 30, 2023, https://www.highnorthnews.com/en/lacking-own-satellite-coverage-russia-looking-china-northern-sea-route-data#:~:text=Russian%20officials%20announced%20a%20host,the%20construction%20of%20future%20icebreakers.

Notes

71 Adam Lajeunesse, "Here There Be Dragons? Chinese Submarine Options in the Arctic," *Journal of Strategic Studies* 6, no. 7 (2022): 1046–62.

72 Thomas Nilsen, "FSB Signs Maritime Security Cooperation with China in Murmansk," Barents Observer (news site), April 25, 2023, https://thebarentsobserver.com/en/security/2023/04/fsb-signs-maritime-security-cooperation-china-murmansk.

73 Wahden, "Big Words, Small Deeds."

74 Moscow Times, "India, Russia Agree Biggest-Ever Oil Deal—Reuters," *Moscow Times*, December 13, 2024, https://www.themoscowtimes.com/2024/12/13/india-russia-agree-biggest-ever-oil-deal-reuters-a87319.

75 Gosudarstvennaia kommissiia po voprosam razvitiia Arktiki, "Rossiia i Indiia dogorovilis' o sotrudnichestve na Dal'nem Vostoke i v Arktike," Gosudarstvennaia kommissiia po voprosam razvitiia Arktiki July 10, 2024, https://arctic.gov.ru/2024/07/0910/%d1%80%d0%be%d1%81%d1%81%d0%b8%d1%8f-%d0%b8-%d0%b8%d0%bd%d0%b4%d0%b8%d1%8f-%d0%b4%d0%be%d0%b3%d0%be%d0%b2%d0%be%d1%80%d0%b8%d0%bb%d0%b8%d1%81%d1%8c-%d0%be-%d1%81%d0%be%d1%82%d1%80%d1%83%d0%b4%d0%bd/.

76 Investment Portal of the Arctic Zone of the Russian Federation, "Indian Icebreakers in the Arctic: Cooperation for the Coming Decades," Arctic-Russia.ru, December 13, 2024, https://arctic-russia.ru/en/article/indian-icebreakers-in-the-arctic-cooperation-for-the-coming-decades/; Raj Kumar Sharma, "The Arctic: The Next Frontier in India-Russia Relations," interview by Natalia Viakhireva, Russian Council (RIAC), June 21, 2024, https://russiancouncil.ru/en/analytics-and-comments/interview/the-arctic-the-next-frontier-in-india-russia-relations/.

77 Heiner Kubny, "Russia and India Plan to Intensify Research in the Arctic," *Polar Journal*, May 17, 2024, https://polarjournal.ch/en/2024/05/17/russia-and-india-plan-to-intensify-research-in-the-arctic//.

78 Kubny, "Russia and India Plan to Intensify Research in the Arctic."

79 Malte Humpert, "Russia Inks Deal with Dubai's DP World to Develop Arctic Container Shipping," High North News, October 26, 2023, https://www.highnorthnews.com/en/russia-inks-deal-dubais-dp-world-develop-arctic-container-shipping.

80 Arctic Russia, "Na samite BRIKS u Rosatoma poiavilis' novye zakazachiki na plavuchie AES," ArcticRussia.ru, November 6, 2024, https://arctic-russia.ru/news/na-sammite-briks-u-rosatoma-poyavilis-novye-zakazchiki-na-plavuchie-aes/.

81 Sergey Vakulenko, "Is a 'Shadow Fleet' of Oil Tankers Really Circumventing the Russian Price Cap?" Carnegie Endowment for International Peace, September 27, 2024, https://carnegieendowment.org/russia-eurasia/politika/2024/09/russia-oil-fleet-sanctions?lang=en&center=russia-eurasia.

82 Thomas Nilsen, "Isolated Russia Invites Faraway Countries to Upcoming Svalbard Science Center in Pyramiden," Barents Observer (news site), October 30, 2023, https://thebarentsobserver.com/en/arctic/2023/10/ghost-town-pyramiden-will-be-home-russias-planned-international-svalbard-science.

Chapter 3

1 Sergei I. Krents, *Itogi ekonomicheskogo razvitiia Arktiki* (Moscow: Institut gosudarstvennogo upravleniia), November 22, 2024, https://igumt.ast.social/menu-news/81-igu031.html.

2 For more on this, see Gustafson, *Klimat*.

3 Alexandre Piliasov, *Arkticheskaia promyshlennost' i promyshlennaia politika* (Moscow, Smolensk: Universum, 2023).

4 For more on this, see Marlene Laruelle and Jean Radvanyi, *Russia: Great Power, Weakened State* (Lanham, MD: Rowman & Littlefield, 2023), 134–7.

5 More in Radvanyi, "When Putin Turned Geographer."

6 Robert G. Jensen, Theodore Shabad, and Arthur W. Wright, eds., *Soviet Natural Resources in the World Economy* (Chicago: University of Chicago Press, 1983).

7 See the seminal work on the costs of exploiting Siberia during the Soviet era: Fiona Hill and Clifford Gaddy, *Siberian Curse: How Communist Planners Left Russia Out in the Cold* (Washington, DC: Brookings Institution, 2003).

8 Fosagro, "AO Apatity prodolzhaet modernizatsiiu obogatelnogo proizvodstva," *Phosagro.ru*, November 10, 2016, https://www.phosagro.ru/press/industry/ao-apatit-prodolzhaet-modernizatsiyu-obogatitelnogo-proizvodstva/.

9 Marlene Laruelle, "Biography of a Polar City: Population Flows and Urban Identity in Norilsk," *Polar Geography* 40, no 4 (2017): 306–23.

10 Natalia Galtseva, "Zolotodobyvaiushchaia otrasl' Magadanskoi oblasti v usloviiakh sanktsii: riski dlia regiona," *EKO* 52, no. 12 (2022): 146–58.

11 Galtseva, "Zolotodobyvaiushchaia otrasl' Magadanskoi oblasti v usloviiakh sanktsii: riski dlia regiona," 151.

Notes

12 Jean Radvanyi, "Réseaux de transport, réseaux d'influence: Nouveaux enjeux stratégiques autour de la Russie," *Hérodote*, no. 104 (2002): 38–65. No volume number.

13 Bojan Pancevski, "A Drunken Evening, a Rented Yacht: The Real Story of the Nord Stream Pipeline Sabotage," *Wall Street Journal*, August 14, 2024, https://www.wsj.com/world/europe/nord-stream-pipeline-explosion-real-story-da24839c?mod=hp_lead_pos7.

14 Valerii Kryukov, Tatyana Skufinoy, and Elena Korchak, eds., *Ekonomika sovremennoi Arktiki: v ostove uspeshnosti, effektivnoe vzaimodeistvie i upravlenie integral'nymi riskami* (Apatity: Izd. Kol'skogo Nauchnogo Tsentra RAN, 2020), 84.

15 Richard Youngs, "A New Geopolitics of EU Energy Security: Vanguardia Dossier," Carnegie Endowment for International Peace, September 23, 2014, https://carnegieendowment.org/posts/2014/09/a-new-geopolitics-of-eu-energy-security?lang=en&center=europe.

16 James Henderson, "Gazprom's LNG Offensive: A Demonstration of Monopoly Strength or Impetus for Russian Gas Sector Reform?" *Post-Communist Economies* 28 (2016): 281–99.

17 Government of the Russian Federation, "Rasporiazhenie Pravitel'stva Rossiiskoi Federatsii of 11.10.2020 no. 1713-r. 'Kompleksnyi plan po razvitiiu proizvodstva szhizhennogo prirodnogo gaza na poluostrove Iamal,'" Government.ru, October 11, 2020, http://government.ru/docs/all/74290/.

18 Riviera, "Winter Season Tests the Mettle of Yamal and Baltic LNG Carriers," Riviera Newsletter, January 22, 2019, https://www.rivieramm.com/opinion/opinion/winter-season-tests-the-mettle-of-yamal-and-baltic-lng-carriers-22109.

19 Dmitry Yakovenko, "Gazovoe siianie: kak Novatek postroil zavod na kraiu zemli," *Forbes Russia*, December 27, 2018, https://www.forbes.ru/milliardery/370603-gazovoe-siyanie-kak-novatek-postroil-zavod-na-krayu-zemli.

20 Julia Louppova, "Russian Arctic Port Sabetta Dispatches First LNG," *Port Today*, December 11, 2017, https://port.today/sabetta-first-lng-dispatch/.

21 Anna Shiryaevskaya and Stephen Bierman, "Putin Calls to Phase Out Gazprom Monopoly on LNG Export," *Bloomberg*, February 13, 2013, http://www.bloomberg.com/news/articles/2013-02-12/putin-set-to-discuss-endinggazprom-s-monopoly-on-lng-exports; Reuters, "Putin Signals End to Gazprom's Russian Gas Export Monopoly," Reuters, June 21, 2013, http://www.reuters.com/article/2013/06/21/us-putin-gas-exports-idUSBRE95K0P220130621, http://www.reuters.com/article/2013/06/21/us-putin-gas-exports-idUSBRE95K0P220130621.

22 Kommersant, "Rossiiskii shel'f stanet amerikanskim," *Kommersant*, February 14, 2013, http://www.kommersant.ru/doc/2126772.

23 Winward, "UPDATED: Illuminating Russia's Shadow Fleet," winward.ai, 2024, https://windward.ai/knowledge-base/illuminating-russias-shadow-fleet/.

24 Julien Bouissou et al., "Russia's Ghost Fleets, a Strategic Asset for Selling Sanctioned Oil," *Le Monde*, October 30, 2024, https://www.lemonde.fr/en/les-decodeurs/article/2024/10/30/russia-s-ghost-fleets-a-strategic-asset-for-selling-sanctioned-oil_6731039_8.html.

25 Oleg S. Anashkin, "Chto nuzhno izmenit' v strategii razvitiia neftepererabatyvaiushchei ostrasli Rossii," *EKO* 52, no. 11 (2022): 158–82.

26 Elizabet Borisenko, "Storony gotovy k nachalu stroitel'stva gazoprovoda Soiuz-Vostokm," *Izvestiia*, September 23, 2024, https://iz.ru/1762326/elizaveta-borisenko/storony-gotovy-k-nachalu-stroitelstva-gazoprovoda-soiuz-vostok; Anton Barbashin, "Putin after Mongolia," *Riddle*, September 17, 2024, https://ridl.io/putin-after-mongolia/.

27 Adrien Sénécat, "L'ambiguïté de Bercy face au projet gazier russe ArcticLNG 2," *Le Monde*, January 25, 2024, https://www.lemonde.fr/les-decodeurs/article/2024/01/24/arctic-lng-2-l-ambiguite-de-bercy-face-au-projet-gazier-russe-sous-sanction-europeenne_6212656_4355770.html.

28 Marine Godelier, "Pourquoi TotalEnergies ne cède pas ses actifs en Russie," *La Tribune*, March 14, 2024, https://www.latribune.fr/climat/energie-environnement/pourquoi-totalenergies-ne-cede-pas-ses-actifs-russes-992859.html.

29 Koen Verhelst, Victor Jack, Antonia Zimmermann, Camille Gijs, and Jürgen Klöckner, "Germany Blocks First-Ever Sanctions on Russian Gas," *Politico*, June 14, 2024., https://www.politico.eu/article/germany-blocks-first-ever-sanctions-russian-gas/; Deutsche Welle, "Evrosoiuz iz-za Germanii ne soglasoval 14-й пакет санкций против РФ." DWi paket sanktsii protiv RF," Deutsche Welle, June 15, 2024, https://www.dw.com/ru/evrosouz-izza-germanii-ne-soglasoval-14j-paket-sankcij-protiv-rf/a-69369251.

30 European Commission, "EU Adopts 14th Package of Sanctions against Russia for Its Continued Illegal War against Ukraine, Strengthening Enforcement and Anti-Circumvention Measures," European Commission, June 24, 2024, https://ec.europa.eu/commission/presscorner/detail/en/ip_24_3423.

31 Malte Humpert, "Russia Still Second Largest Gas Provider to EU, after Norway, with LNG Imports Increasing," High North News, September

Notes

12, 2024, https://www.highnorthnews.com/en/russia-still-second-largest-gas-provider-eu-after-norway-lng-imports-increasing.

32 Jean Radvanyi, "Les ressources naturelles russes à l'heure de la guerre en Ukraine: Un atout empoisonné?" *Hérodote*, no. 190–1 (2023): 119–36. No volume number.

33 Kryukov, Skufinoy, and Korchak, *Ekonomika sovremennoi Arktiki*, 13–14.

34 Robert Gres, "Arkticheskaia spetsifika: Kontent-analiz strategii regionov i munitsipalitetov rossiiskoi arktiki," *Regional'noe issledovanie* 83, no. 1 (2024): 88–100.

35 Natalia Zubarevich, Sergey Artobolevsky, Olga Vendina, Oleg Gonmakher, and Aleksandr Kynev, "Ob"edenenie sub"ektov rossiskoi federatsii: Za i protiv," *Lobbyist* 24, no. 6 (2010), 24–38, http://emsu.ru›extra/pdf5s/lobbyist/2010/6/24.pdf.

36 Kryukov, Skufinoy, and Korchak, *Ekonomika sovremennoi Arktiki*, 20.

37 Government of the Russian Federation, "Postanovlenie Pravitelstva 'Ob utverzhdenii polozheniia o Gosudarstvennoi komissii po voprosam razvitiarazvitiia Arktiki,'" *Government.ru*, March 14, 2015, http://government.ru/docs/all/95235/http://government.ru/docs/all/95235/.

38 Artur Kuchumov, Elena Pecheritsa, and Natalia Blazhenkova, "Problems of Entrepreneurship Development in the Arctic: Russian and Foreign Experience of Observing the Principles of Green Economy," *E3S Web of Conferences* 378, 06002 (2023): 1–6, www.e3s-conferences.org/articles/e3sconf/pdf/2023/15/e3sconf iirpcmia2023_06002.pdf.

39 Neftegaz, "Platforma Prirazlomnaya," Neftegaz.ru, March 27, 2015, https://neftegaz.ru/science/booty/331727-platforma-prirazlomnaya-element-uverennosti-i-moshchi-rossii/.

40 Charles Digges, "Russia's Giant Shtokman Gas Field Project Put on Indefinite Hold over Cost Overruns and Failed Agreements," Bellona.org, August 29, 2012, https://bellona.org/news/fossil-fuels/gas/2012-08-russias-giant-shtokman-gas-field-project-put-on-indefinite-hold-over-cost-overruns-and-failed-agreements.

41 Mining Technologies, "Kinross Divests Russian Assets to Highland Gold at Half the Set Amount," Mining-technology.com, June 16, 2022, https://www.mining-technology.com/news/kinross-russian-assets-highland/.

42 Kryukov, Skufinoy, and Korchak, *Ekonomika sovremennoi Arktiki*, 17.

43 The Hectare program has been in place in the Far East since 2016. Since then, 152,000 Russian citizens have benefited from it. It has been extended until 2035. https://www.cian.ru/stati-dalnevostochnyj-gektar-kak-poluchit-uchastok-ot-gosudarstva-v-2025-godu-339574/?ysclid=mcx70r6980127111374.

44 Natalya Shapovalova, "Kak vyrashivaiut ovoshchi v regione nevozmozhnogo zemledeliia," *AgroXXI.ru*, December 7, 2021, https://www.agroxxi.ru/gazeta-zaschita-rastenii/zrast/kak-vyraschivayut-ovoschi-v-regione-nevozmozhnogo-zemledelija.html.

45 Vladimir Tishak, Enver Ibragimov, and Ivan Nosov, "Osobennosti arkticheskogo teplichnogo sel'skogo khoziaistvoveniia," goartic.ru, September 9, 2021, https://goarctic.ru/news/osobennosti-arkticheskogo-teplichnogo-selskogo-khozyaystvovaniya/?ysclid=m418crsdxp784261339.

46 Kryukov, Skufinoy, and Korchak, *Ekonomika sovremennoi Arktiki*, 122.

47 Josh Gabbatiss, "Arctic Ocean Fishing Ban Welcomed by Scientists and Environmentalists," December 4, 2017, *The Independent*, https://www.independent.co.uk/climate-change/news/arctic-ocean-fishing-ban-environment-scientists-populations-fish-a8091541.html.

48 Dimitry N. Ostrovskii, ed., *Putevoditel' po Severu Rossii* (Saint Petersburg: Izd. Tovarishchestva Arkhangel'sko-Murmanskogo parokhodstva, 1898); Yury F. Lukin, *Arkticheskii turizm v Rossii* (Arkhangelsk: SAFU, 2016).

49 Valeriia Kruzhalin, Natalia Shabalina, Ekaterian Kashirina, and Alexandra Nikanorova, "Specific Issues of Arctic Cruise Development in the Russian Arctic: The Russian Arctic National Park Case Study," *Anais Brasileiros de Estudos Turísticos* 12 (2022): 1–10; Pierre-Louis Têtu, Jackie Dawson, and Frédéric Lasserre, "The Evolution and Relative Competitiveness of Global Arctic Cruise Tourism Destinations," in *Arctic Shipping. Climate Change, Commercial Traffic and Port Development*, ed. Frédéric Lasserre and Olivier Faury (London: Routledge, 2020), 94–114.

50 Galia Morell, "Barneo Is on This Year, But Do You Want to Go?" Explorersweb.com, March 14, 2024, https://explorersweb.com/barneo-do-you-want-to-go/.

51 Arctic Russia, "Arkticheskie kruizy: Ekskliuzivnyi draiver rosta SMP," Arctic.russia.ru, March 3, 2024, https://arctic-russia.ru/article/arkticheskie-kruizy-eksklyuzivnyy-drayver-rosta-smp/.

52 1prime.ru, "Antirossiiskie sanktsii narushili torgovliu krupneishego almaznogo tsentra v ES," 1prime.ru, March 19, 2024, https://1prime.ru/20240319/diamonds-846485741.html?ysclid=m0zd55z0tg614819545.

53 Alexander Piliasov, professor of geography at Lomonosov Moscow State University and the Higher School of Economics in Moscow, interview by Jean Radvanyi, January 2024.

54 Kryukov, Skufinoy, and Korchak, *Ekonomika sovremennoi Arktiki*, 127.

Notes

Chapter 4

1. On the multiplicity of Arctic identities and ways of life, see Hohmann Sophie and Dominique Samson Normand de Chambourg, eds., "Jardins d'hiver: Paysages culturels du Nord et de l'Arctique sibériens," *Slavica Occitania*, no. 58 (2024): 477 pages.
2. More details in Terence Armstrong, *Russian Settlement in the North* (Cambridge: Cambridge University Press, 1965).
3. John McCannon, *Red Arctic: Polar Exploration and the Myth of the North in the Soviet Union, 1932–1939* (New York: Oxford University Press, 1998), 256.
4. McCannon, *Red Arctic*; see also Piers Horensma, *The Soviet Arctic* (London: Routledge).
5. Peter Hemmersam, *Making the Arctic City: The History and Future of Urbanism in the Circumpolar North* (London: Bloomsbury, 2021); P. A., Filin, M. A. Emelina, and M. A. Savinov, *Arktika za gran'iu fantastiki: Budushchee Severa glazami sovetskikh inzhenerov, izobretatelei i pisatelei* (Moscow: Paulsen, 2018).
6. On Russian geographical metanarratives, see Marlene Laruelle, "Larger, Higher, Farther North … Geographical Metanarratives of the Nation in Russia," *Eurasian Geography and Economics* 53, no. 5 (2012): 557–74.
7. Aleksandr Prokhanov, "Ledovityi okean—vnutrennee more Rossii," *Zavtra*, August 8, 2007, http://www.zavtra.ru/cgi/veil/data/zavtra/07/716/11.html.
8. Aleksandr Prokhanov, "Severnyi marsh Rossii," *Zavtra*, April 25, 2007, http://www.arctictoday.ru/analytics/704.html.
9. Marlene Laruelle, "The Three Waves of Arctic Urbanization: Drivers, Evolutions, Prospects," *Polar Record* 55, no. 1 (2019): 1–12.
10. V. V. Fauzer, A. V. Smirnov, T. S. Lytkina, and G. N. Fauzer, *Rossiiskaia i mirovaia Arktika: naselenie, ekonomika, rasselenie. Monografiia* (Moscow: Rosspen, 2022, 215 pages); Nadezhda Iu. Zamiatina i R. V. Goncharov, "Arkticheskaia urbanizatsiia: fenomen i sravnitel'nyi analiz," *Vestnik Moskovskogo Universiteta. Geografiia*, no. 5 (2020): 69–82; Nadezhda Iu. Zamiatina, *Zhiznestoikost' arkticheskikh gorodov: Teoriia, kompleksnyi analiz i primery transformatsii* (Moscow: Izdatel'skie resheniia, 2023); Nadezhda Iu. Zamiatina, "Development Cycles of Cities in the Siberian North," in *The Siberian World*, ed. John P. Ziker, Jenanne Ferguson, and Vladimir Davydov (London: Routledge, 2023), 352–63.
11. Marlene Laruelle, "Urban Regimes in Russia's Northern Cities: Testing a Concept in a New Environment," *Arctic* 73, no. 1 (2020): 53–66.

Notes

12 Gavin Slade, Laura Piacentini, and Alena Kravtsova, "Ghosts of the Gulag: Negotiating Spectres of the Penal Past in Northern Russia," *British Journal of Criminology* 64, no. 1 (2024): 17–33; see also Tyler C. Kirk, *After the Gulag: A History of Memory in Russia's Far North* (Bloomington: Indiana University Press, 2023).

13 Tatiana Mikhailova, "Gulag, WWII and the Long-Run Patterns of Soviet City Growth," *Munich Personal RePEc Archive Papers*, September 2012, https://econpapers.repec.org/paper/pramprapa/41758.htm.

14 On international patterns of Arctic urbanization, see Laruelle, "The Three Waves of Arctic Urbanization."

15 Natal'ia Zubarevich, "Four Russias: Human Potential and Social Differentiation of Russian Regions and Cities," in *Russia 2025*, ed. Maria Lipman and Nikolai Petrov (London: Palgrave Macmillan, 2015), 67–85.

16 Nadezhda Iu. Zamiatina, "Igarka as a Frontier: Lessons from the Pioneer of the Northern Sea Route," *Journal of Siberian Federal University* 13, no. 5 (2020): 783–99.

17 Nikolai Bobylev, Sebastien Gadal, Valery Konyshev, Maria Lagutina, and Alexander Sergunin, "Building Urban Climate Change Adaptation Strategies: The Case of Russian Arctic Cities," *Weather, Climate, and Society* 13, no. 4 (2021): 875–84.

18 Irina Busygina, "The Role of Russia's Regions in the War Economy," Zentrum für Osteuropa-und internationale Studien (Berlin), ZOiS Spotlight 22/2024, November 27, 2024, https://www.zois-berlin.de/en/publications/zois-spotlight/the-role-of-russias-regions-in-the-war-economy.

19 Yulia Kuzmina, "'The Defenders of Shiyes': Traditionalism as a Mobilisation Resource in a Russian Protest Camp," *East European Politics* 39, no. 2 (2022): 260–80.

20 Niobe Thompson, "Administrative Resettlement and the Pursuit of Economy: The Case of Chukotka," *Polar Geography* 26, no. 4 (2002): 270–88. See also the project funded by the BOREAS program of the European Science Foundation, "Moved by the State: Perspectives on Relocation and Resettlement in the Circumpolar North (MOVE), https://www.sciencedaily.com/releases/2007/04/070405180552.htm.

21 Timothy Heleniak, "The Role of Attachment to Place in Migration Decisions of the Population of the Russian North," *Polar Geography* 32, nos. 1–2 (2009): 31–60.

22 Alla Bolotova, Anastasia Karaseva, and Valeria Vasilyeva, "Mobility and Sense of Place among Youth in the Russian Arctic," *Sibirica: Journal of Siberian Studies* 16, no. 3 (2017): 77–123.

Notes

23 Heleniak, "The Role of Attachment to Place in Migration Decisions of the Population of the Russian North."

24 Nikolay Shiklomanov, Dmitry Streletskiy, Luis Suter, Robert Orttung, and Nadezhda Zamyatina, "Dealing with the Bust in Vorkuta, Russia," *Land Use Policy* 93 (2020): 1–11, https://par.nsf.gov/servlets/purl/10161894.

25 Svetlana Sukneva and Marlene Laruelle, "A Booming City in the Far North: Demographic and Migration Dynamics of Yakutsk, Russia," *Sibirica: The Journal of Siberian Studies* 9, no. 3 (2019): 9–28.

26 Nadezhda Zamyatina et al., "Shrinking Cities, Growing Cities: A Comparative Analysis of Vorkuta and Salekhard," in *Urban Sustainability in the Arctic: Measuring Progress in Circumpolar Cities*, ed. Robert W. Orttung (New York: Berghahn Books, 2020), 49–73.

27 Timothy Heleniak, "Polar Peoples in the Future: Projections of the Arctic Populations," *Nordregio Working Paper* no. 3 (2020), https://nordregio.org/publications/polar-peoples-in-the-future-projections-of-the-arctic-populations/.

28 Nadezhda Zamiatina and Ruslan Goncharov, "'Agglomeration of Flows': Case of Migration Ties between the Arctic and the Southern Regions of Russia," *Regional Science Policy & Practice* 14, no. 1 (2022): 63.

29 Nadezhda Zamiatina and Ruslan Goncharov, "Population Mobility and the Contrasts between Cities in the Russian Arctic and Their Southern Russian Counterparts," *Area Development and Policy*, Vol. 3 no. 3 (2018): 293–308.

30 Sergei Ermolaev, "The Formation and Evolution of the Soviet Union's Oil and Gas Dependence," Carnegie Endowment for International Peace Working Paper, March 29, 2017, https://carnegieendowment.org/posts/2017/03/the-formation-and-evolution-of-the-soviet-union-s-oil-and-gas-dependence?lang=en.

31 This is a summary of Marlene Laruelle and Sophie Hohmann, "Introduction" to the Special Issue "Polar Islam: The Rise of Islam in Russia's Far North," *Problems of Post-Communism* 67, nos. 4–5 (2019): 1–9. For an overview of migration in Russia in the early 2010s, see Sergei Guriev and Elena Vakulenko, "Breaking Out of Poverty Traps: Internal Migration and Interregional Convergence in Russia," *Journal of Comparative Economics* 43, no. 3 (2015): 633–49; and Mikhail A. Alexseev, "Russia and Central Asia," in *Handbook on Migration and Security*, ed. Philippe Bourbeau (Northampton, MA: Edward Elgar, 2017), 263–94.

32. United Nations, "Internal Migration Report 2013," UN.org, 2013, http:// www.un.org/en/development/desa/population/publications/pdf/migration/migrationreport2013/Full_Document_final.pdf.
33. Local geologist working for one of the main mining companies of the region, interviewed by Marlene Laruelle and Sophie Hohmann, Apatity, July 2015.
34. Nikolai Mkrtchian and Liliya Karachurina, "Migratsiia v Rossii: Potoki i tentry pritiazheniia," *Demoscope Weekly* nos. 595–596 (April 21–May 4), 2014, https://www.demoscope.ru/weekly/2014/0595/demoscope 595.pdf.
35. Denis Sokolov, "Ugra, the Dagestani North: Anthropology of Mobility between the North Caucasus and Western Siberia," in *New Mobilities and Social Changes in Russia's Arctic Regions*, ed. Marlene Laruelle (Abingdon: Routledge, 2016), 176–93; Ahmet Yarlykapov, "Divisions and Unity of the Novy Urengoy Muslim Community," *Problems of Post-Communism* 67, nos. 4–5 (2020): 338–47.
36. Sophie Roche, "Illegal Migrants and Pious Muslims: The Paradox of Bazaar Workers from Tajikistan," in *Tajikistan on the Move: Statebuilding and Societal Transformations*, ed. Marlene Laruelle (Lanham, MD: Lexington Books, 2018), 247–78.
37. Marlene Laruelle and Sophie Hohmann, "Polar Islam: Muslim Communities in Russia's Arctic Cities," in "Polar Islam: The Rise of Islam in Russia's Far North," ed. Marlene Laruelle and Sophie Hohmann, *Problems of Post-Communism* 67, no. 4–5 (2019): 327–37.
38. Sokolov, "Ugra, the Dagestani North: Anthropology of Mobility between the North Caucasus and Western Siberia," 176–93; see also Denis Sokolov, "Islam in the Gold Heart of Russia: Ingush Islamic Communities in the Kolyma," *Problems of Post-Communism* 67, nos. 4–5 (2019): 348–61.
39. Florian Vidal, "Russia in the Arctic: The End of Illusions and the Emergence of Strategic Realignments," Institut Français des Relations Internationales (IFRI), July 2024, https://www.ifri.org/en/papers/russia-arctic-end-illusions-and-emergence-strategic-realignments.
40. Nordregio, "Indigenous Population in the Arctic," Nordregio.org, March 2019, https://www.nordregio.org/maps/indigenous-population-in-the-arctic/.
41. Adam Lenton et al., *Decolonizing Russia? Disentangling Debates* (Cambridge: Cambridge University Press, 2025); Andreas Kappeler, *The Russian Empire: A Multi-Ethnic History*, 1st ed. (London: Routledge, 2001).

Notes

42 Slezkine, *Arctic Mirrors*; Terry Martin, *The Affirmative Action Empire: Nations and Nationalism in the USSR, 1923–1939* (Ithaca, NY: Cornell University Press, 2001).

43 Indra Overland and Helge Blakkisrud, "The Evolution of Federal Indigenous Policy in the Post-Soviet North," in *Tackling Space: Federal Politics and the Russian North*, ed. Helge Blakkisrud and Geir Hønneland (Lanham, MD: University Press of America, 2006), 175–6.

44 Arctic Council Indigenous Peoples Secretariat, "Language Revitalization: Nomadic Schools in the Republic of Sakha (Yakutia), Russia," Arctic Council Indigenous Peoples Secretariat, 2019, https://www.arcticpeoples.com/sagastallamin-revitalization-nomadic-schools-sakha.

45 Data from Rosstat, Russia's 2021 population census, https://eng.rosstat.gov.ru/folder/76215.

46 Ryan Weber et al., "Urbanization and Land Use Management in the Arctic: An Investigative Overview," in *Northern Sustainabilities: Understanding and Addressing Change in the Circumpolar World*, ed. Gail Fondahl and Gary N. Wilson (New York: Springer, 2017), 269–84.

47 For more on this, see Violetta Gassiy, "Protecting Indigenous Rights from Mining Companies: The Case of Ethnological Expertise in Yakutia," *Sibirica: Interdisciplinary Journal of Siberian Studies* 18, no. 3 (2019): 92–108.

48 T. N. Vasil'kova, A. V. Evay, E. P. Martynova, and N. I. Novikova, *Korennye malochislennye narody i promyshlennoe razvitie Arktiki* (Moscow: Shchadrinskii dom pechati, 2011); E. P. Martynova and N. I. Novikova, *Tazovskie nentsy v usloviiakh neftegazovogo osvoeniia* (Moscow: RAN, 2012).

49 Gassiy, "Protecting Indigenous Rights from Mining Companies."

50 Yurii Simchenko, "Narody severa Rossii," *Issledovaniia po prikladnoi i neotlozhnoi etnologii* no. 112 (1998): 1–26.

51 Alexandra Tomaselli and Anna Koch, "Implementation of Indigenous Rights in Russia: Shortcomings and Recent Developments," *International Indigenous Policy Journal* 5, no. 4 (2014): 1–21; Ron Wallace, "The Case for RAIPON: Implications for Canada and the Arctic Council" (Calgary, Alberta: Canadian Defense and Foreign Affairs Institute, 2013, http://www.iwgia.org/iwgia_files_publications_files/0695_HumanRights_report_18_Russia.pdf.

52 Andrey Plotnitskiy and Arnab Roy Chowdhury, "Killing Nature—Killing Us: 'Cultural Threats' as a Fundamental Framework for Analyzing Indigenous Movements against Mining in Siberia and the Russian North," *Post-Soviet Affairs* 39, no. 3 (2023): 195–212.

53 Ekaterina Zmyvalova, "The Rights of Indigenous Peoples of Russia after Partial Military Mobilization," *Arctic Review on Law and Politics* 14 (2023): 70–5.
54 Alexsey Bessudnov, "Ethnic and Regional Inequalities in Russian Military Fatalities in Ukraine: Preliminary Findings from Crowdsourced Data," *Demographic Research* 48 (2023): 883–98.
55 Timothy Heleniak, "Arctic Populations and Migrations," in *Arctic Human Development Report: Regional Processes and Global Linkages*, ed. Joan Nyman Larsen, Joan Nymand, and Gail Fondahl (Oslo: Nordic Council of Ministers, 2015), 53–104.
56 Laur Vallikivi, *Words and Silences: Nenets Reindeer Herders and Russian Evangelical Missionaries in the Post-Soviet Arctic* (Bloomington: Indiana University Press, 2024).
57 Marya Rozanova, "Indigenous Urbanization in Russia's Arctic: The Case of Nenets Autonomous Region," *Sibirica: Interdisciplinary Journal of Siberian Studies* 18, no. 3 (2019): 54–91; Mikkel Berg-Nordlie, Astri Dankertsen, and Marte Winsvold, eds., *An Urban Future for Sapmi? Indigenous Urbanization in the Nordic States and Russia* (New York: Berghahn, 2022); Lenore A. Grenoble, "Contact and Shift: Colonization and Urbanization in the Arctic," in *The Cambridge Handbook of Language Contact*, ed. Salikoko Mufwene and Anna Maria Escobar (Cambridge: Cambridge University Press, 2022), 473–501.
58 Marlene Laruelle, "Postcolonial Polar Cities? New Indigenous and Cosmopolitan Urbanness in the Arctic," *Acta Borealia* 36, no. 2 (2019): 149–65.
59 Marya Rozanova-Smith, Stanislav Ksenofontov, and Andrey N. Petrov, "Indigenous Urbanization and Indigenous Urban Experiences in the Russian Arctic: The Cases of Yakutsk and Naryan-Mar," in *Urban Indigeneities*, ed. Dana Brablec and Andrew Canessa (Tucson: University of Arizona Press, 2023), 209–42.
60 Vera Kuklina, Sargylana Ignatieva, and Uliana Vinokurova, "Educational Institutions as a Resource for the Urbanization of Indigenous People: The Case of Yakutsk," *Sibirica: Journal of Siberian Studies* 8, no. 3 (2019): 29–53; Jenanne Ferguson and Lena Sidorova, "What Language Advertises: Ethnographic Branding in the Linguistic Landscape of Yakutsk," *Language Policy* 17, no. 1 (2018): 23–54.

Notes

Conclusion

1 Timo Koivurova and Akiho Shibata, "After Russia's Invasion of Ukraine in 2022: Can We Still Cooperate with Russia in the Arctic?" *Polar Record* 59, no. 12 (2023): 1–9; Stefan Kirchner, "International Arctic Governance without Russia," *SSRN* (2022): 1–9, https://ssrn.com/abstract=4044107; Alexandra Witze, "Russia's War in Ukraine Forces Arctic Climate Projects to Pivot," *Nature*, July 11, 2020, https://www.nature.com/articles/d41586-022-01868-9.

2 On Arctic urban sustainability, see Robert Orttung, ed., *Urban Sustainability in the Arctic: Measuring Progress in Circumpolar Cities* (New York: Berghahn Books, 2020) ; Robert Orttung, ed., *Sustaining Russia's Arctic Cities: Resource Politics, Migration, and Climate Change* (New York: Berghahn Books, 2016).

3 Andrey N. Petrov, Marya S. Rozanova, Andrey K. Krivorotov, Elena M. Klyuchnikova, Valeriy L. Mikheev, Alexander N. Pelyasov, and Nadezhda Yu. Zamyatina, "The Russian Arctic by 2050: Developing Integrated Scenarios," *Arctic* 74, no. 3 (2021): 306–22.

SELECTED BIBLIOGRAPHY

Akimov, Yury. "Arctic Paradiplomacy of the Republic of Sakha (Yakutia): The Impact of Federalism, Nationalism, and Identity." In *Mapping Arctic Paradiplomacy. Limits and Opportunities for Sub-National Actors in Arctic Governance*, edited by Mathieu Landriault, Jean-François Payette, and Stéphane Roussel, 77–98. London: Routledge, 2022.

Anisimov, Oleg, and Svetlana Reneva. "Permafrost and Changing Climate: The Russian Perspective." *Ambio* 35, no. 4 (2006): 169–75.

Armstrong, Terence. *The Northern Sea Route: Soviet Exploitation of the North East Passage*. Cambridge: Cambridge University Press, 2011.

Armstrong, Terence. *Russian Settlement in the North*. Cambridge: Cambridge University Press, 1965.

Arzyutov, Dmitry. "The Making of the *Homo Polaris*: Human Acclimatization to the Arctic Environment and Soviet Ideologies in Northern Medical Institutions." *Settler Colonial Studies* 14, no. 2 (2023): 180–203.

Åtland, Kristian. "Mikhail Gorbachev, the Murmansk Initiative, and the Desecuritization of Interstate Relations in the Arctic." *Cooperation and Conflict* 43, no. 3 (2008): 289–311.

Åtland, Kristian, and Torbjørn Pedersen. "The Svalbard Archipelago in Russian Security Policy: Overcoming the Legacy of Fear—or Reproducing It?" *European Security* 17, no. 2 (2008): 227–51.

Baev, Pavel K. "Russia's Ambivalent Status-Quo/Revisionist Policies in the Arctic." *Arctic Review on Law and Politics* 9 (2018): 408–24.

Baev, Pavel K. "Threat Assessments and Strategic Objectives in Russia's Arctic Policy." *The Journal of Slavic Military Studies* 32, no. 1 (2019): 25–40.

Bennett, Mia M., Scott R. Stephenson, Kang Yang, Michael T. Bravo, and Bert De Jonghe. "The Opening of the Transpolar Sea Route: Logistical, Geopolitical, Environmental, and Socioeconomic Impacts." *Marine Policy* 121 (2020): 1–15.

Bobylev, Nikolai, Sebastien Gadal, Valery Konyshev, Maria Lagutina, and Alexander Sergunin. "Building Urban Climate Change Adaptation Strategies: The Case of Russian Arctic Cities." *Weather, Climate, and Society* 13, no. 4 (2021): 875–84.

Selected Bibliography

Bolotova, Alla, Anastasia Karaseva, and Valeria Vasilyeva. "Mobility and Sense of Place among Youth in the Russian Arctic." *Sibirica: Journal of Siberian Studies* 16, no. 3 (2017): 77–123.

Boulègue, Mathieu. "Russia's Military Posture in the Arctic: Managing Hard Power in a 'Low Tension' Environment." Chatham House Research Paper, October 15, 2024. https://www.chathamhouse.org/2019/06/russias-military-posture-arctic/3-military-infrastructure-and-logistics-russian-arctic.

Boylan, Brandon M. "Increased Maritime Traffic in the Arctic: Implications for Governance of Arctic Sea Routes." *Marine Policy* 131 (2021): 1–8. https://www.sciencedirect.com/science/article/abs/pii/S0308597X21001779.

Butler, William E. *The Northeast Arctic Passage*. Alphen aan den Rijn: Sijthoff & Noordhoof International, 1978. https://archive.org/details/northeastarcticp0000butl.

Devyatkin, Pavel. "Arctic Exceptionalism: A Narrative of Cooperation and Conflict from Gorbachev to Medvedev and Putin." *Polar Journal* 13, no. 2 (2023): 336–57.

Dunlap, William V. "Transit Passage in the Russian Arctic Straits." *International Boundaries Research Unit Maritime Briefing* 1, no. 7 (1996): 1–79.

Elferink, Alex G. "The Law and Politics of the Maritime Boundary Delimitation of the Russian Federation, Part 1." *International Journal of Marine and Coastal Law* 11, no. 4 (1996): 525–61.

Elferink, Alex G. "The Law and Politics of the Maritime Boundary Delimitation of the Russian Federation, Part 2." *International Journal of Marine and Coastal Law* 12, no. 1 (1997): 5–35.

Esau, Igor, and Victoria Miles. "Warmer Urban Climates for Development of Green Spaces in Northern Siberian Cities." *Geography, Environment, Sustainability* 9, no. 4 (2016): 48–62.

Exner-Pirot, Heather, and Robert W. Murray. "Regional Order in the Arctic: Negotiated Exceptionalism." *Politik* 20, no. 3 (2017): 47–64.

Fauzev, V. V. et al. *Rossiiskaia i mirovaia Arktika: naselenie, ekonomika, rasselenie—Monografiia*. Moscow: Rosspen, 2022.

Fedi, Laurent Etienne, and Olivier Faury. "Mapping and Analysis of Maritime Accidents in the Russian Arctic through the Lens of the Polar Code and POLARIS System." *Marine Policy* 118 (2020): 1–28.

Ferguson, Jenanne, and Lena Sidorova. "What Language Advertises: Ethnographic Branding in the Linguistic Landscape of Yakutsk." *Language Policy* 17, no. 1 (2018): 23–54.

Filin, P. A., M. A. Emelina, and M. A. Savinov. *Arktika za gran'iu fantastiki: Budushchee Severa glazami sovetskikh inzhenerov, izobretatelei i pisatelei*. Moscow: Paulsen, 2018.

Selected Bibliography

Fravel, M. Taylor, Kathryn Lavelle, and Lieselotte Odgaard. "China Engages the Arctic: A Great Power in a Regime Complex." *Asian Security* 18, no. 2 (2021): 138–58.

Galtseva, Natalia. "Zolotodobyvaiushchaia otrasl' Magadanskoi oblasti v usloviiakh sanktsii: riski dlia regiona." *EKO*, no. 12 (2022): 146–58.

Gassiy, Violetta. "Protecting Indigenous Rights from Mining Companies: The Case of Ethnological Expertise in Yakutia." *Sibirica: Interdisciplinary Journal of Siberian Studies* 18, no. 3 (2019): 92–108.

Gres, Robert. "Arkticheskaia spetsifika: Kontent-analiz strategii regionov i munitsipalitetov rossiiskoi arktiki." *Regional'noe issledovanie* no. 1 (2024): 88–100.

Gustafson, Thane. *Klimat: Russia in the Age of Climate Change*. Cambridge, MA: Harvard University Press, 2021.

Heleniak, Timothy. "Arctic Populations and Migrations." In *Arctic Human Development Report: Regional Processes and Global Linkages*, edited by Joan Nymand Larsen and Gail Fondahl, 53–104. Oslo: Nordic Council of Ministers, 2015.

Heleniak, Timothy. "The Role of Attachment to Place in Migration Decisions of the Population of the Russian North." *Polar Geography* 32, nos. 1–2 (2009): 31–60.

Heleniak, Timothy. "Polar Peoples in the Future: Projections of the Arctic Populations." Nordregio Working Paper no. 3 (2020). https://nordregio.org/publications/polar-peoples-in-the-future-projections-of-the-arctic-populations/.

Hemmersam, Peter. *Making the Arctic City: The History and Future of Urbanism in the Circumpolar North*. London: Bloomsbury, 2021.

Henriksen, Tore, and Geir Ulfstein. "Maritime Delimitation in the Arctic: The Barents Sea Treaty." *Ocean Development & International Law* 42, nos. 1–2 (2011): 1–21.

Hill, Fiona, and Clifford Gaddy. *Siberian Curse: How Communist Planners Left Russia Out in the Cold*. Washington, DC: Brookings Institution Press, 2003.

Hohmann, Sophie, and Dominique Samson Normand de Chambourg, eds. "Jardins d'hiver: Paysages culturels du Nord et de l'Arctique sibériens." *Slavica Occitania* no. 58 (2024).

Horensma, Piers. *The Soviet Arctic*. London: Routledge, 1991.

Kirk, Tyler C. *After the Gulag: A History of Memory in Russia's Far North*. Bloomington: Indiana University Press, 2023.

Koivurova, Timo, and Akiho Shibata. "After Russia's Invasion of Ukraine in 2022: Can We Still Cooperate with Russia in the Arctic?" *Polar Record* 59, no. 12 (2023): 1–9.

Selected Bibliography

Konyshev, Valery, and Alexander Sergunin. "Is Russia a Revisionist Military Power in the Arctic?" *Defense & Security Analysis* 30, no. 4 (2014): 323–35.

Kryukov, Valerii T. Skufinoy, and E. Korchak, eds. *Ekonomika sovremennoi Arktiki: V ostove uspeshnosti, effektivnoe vzaimodeistvie i upravlenie integral'nymi riskami*. Apatity: Izdanie Kol'skogo Nauchnogo Tsentra Rossiiskoi Akademii Nauk, 2020.

Kuklina, Vera, Sargylana Ignatieva, and Uliana Vinokurova. "Educational Institutions as a Resource for the Urbanization of Indigenous People: The Case of Yakutsk." *Sibirica: Journal of Siberian Studies* 8, no. 3 (2019): 29–53.

Lanteigne, Marc. "Considering the Arctic as a Security Region: The Roles of China and Russia." In *Routledge Handbook of Arctic Security*, edited by Gunhild Hoogensen Gjørv, Marc Lanteigne, and Horatio Sam-Aggrey, 1–13. London: Routledge, 2017.

Laruelle, Marlene. "Postcolonial Polar Cities? New Indigenous and Cosmopolitan Urbanness in the Arctic." *Acta Borealia* 36, no. 2 (2019): 149–65.

Laruelle, Marlene. *Russia's Arctic Strategies and the Future of the Far North*. London: Routledge, 2013.

Laruelle, Marlene. "The Three Waves of Arctic Urbanization: Drivers, Evolutions, Prospects." *Polar Record* 55, no. 1 (2019): 1–12.

Laruelle, Marlene. "Urban Regimes in Russia's Northern Cities: Testing a Concept in a New Environment." *Arctic* 73, no. 1 (2020): 53–66.

Laruelle, Marlene, Igor Esau, Martin Miles, Victoria Miles, Anna N. Kurchatova, and Sergej A. Petrov. "Arctic Cities as an Anthropogenic Object: A Preliminary Approach through Urban Heat Islands." *Polar Journal* 9, no. 2 (2019): 402–23.

Laruelle, Marlene, and Sophie Hohmann. "Polar Islam: Muslim Communities in Russia's Arctic Cities." *Problems of Post-Communism* 67, nos. 4–5 (2019): 327–37.

Laruelle, Marlene, and Nikolai Shiklomanov. "Special issue on Norilsk." *Polar Geography* 40, no. 4 (2017).

Lavelle, Kathryn C. "Regime, Climate, and Region in Transition: Russian Participation in the Arctic Council." *Problems of Post-Communism* 69, nos. 4–5 (2022): 345–57.

Leksyutina, Yana V. "China's Participation in Energy Cooperation with Russia in the Arctic." In *Energy of the Russian Arctic*, edited by Valery I. Salygin. Berlin: Springer, 2022, 125–40.

Lipunov, Nikita, and Pavel Devyatkin. "The Arctic in the 2023 Russian Foreign Policy Concept." The Arctic Institute, May 30, 2023. https://www.thearcticinstitute.org/arctic-2023-russian-foreign-policy-concept/.

Selected Bibliography

Martus, Ellie. "Policymaking and Policy Framing: Russian Environmental Politics under Putin." *Europe-Asia Studies* 73, no. 5 (2021): 869–89.

Martynova, E. P., and N. I. Novikova. *Tazovskie nentsy v usloviiakh neftegazovogo osvoeniia*. Moscow: Rossiiskaia Akademia Nauk, 2012.

Matz-Lück, Nele. "Planting the Flag in Arctic Waters: Russia's Claim to the North Pole." *Göttingen Journal of International Law* 1, no. 2 (2009): 235–55.

McCannon, John. *Red Arctic: Polar Exploration and the Myth of the North in the Soviet Union, 1932–1939*. New York: Oxford University Press, 1998.

Mikhailova, Tatiana. "Gulag, WWII and the Long-Run Patterns of Soviet City Growth." *Munich Personal RePEc Archive Papers*, September 2012. https://econpapers.repec.org/paper/pramprapa/41758.htm.

Moe, Arild. "The Russian Barents Sea: Openings for Norway?" In *High North: High Stakes*, edited by Rosa Gottemoeller and R. Tamnes, 75–85. Bergen: Fagbokforlaget, 2008.

Moe, Arild, Daniel Fjærtoft, and Indra Øverland. "Space and Timing: Why Was the Barents Sea Delimitation Dispute Resolved in 2010?" *Polar Geography* 34, no. 3 (2011): 145–62.

Nord, Douglas C. *The Arctic Council: Governance within the Far North*. London: Routledge, 2016.

Orttung, Robert, ed. *Sustaining Russia's Arctic Cities: Resource Politics, Migration, and Climate Change*. New York: Berghahn Books, 2016.

Orttung, Robert, ed. *Urban Sustainability in the Arctic: Measuring Progress in Circumpolar Cities*. New York: Berghahn Books, 2020.

Østhagen, Andreas. "High North, Low Politics—Maritime Cooperation with Russia in the Arctic." *Arctic Review on Law and Politics* 7, no. 1 (2016): 83–100.

Overland, Indra, and Helge Blakkisrud. "The Evolution of Federal Indigenous Policy in the Post-Soviet North." In *Tackling Space: Federal Politics and the Russian North*, edited by Helge Blakkisrud and Geir Hønneland, 175–6. Lanham, MD: University Press of America, 2006.

Pedersen, Torbjørn, and Beate Steinveg, "Russia's Clashing Ambitions: Arctic Status Quo and World-Order Revision." *Politics and Governance* 12 (2024): 1–12.

Petrov, Andrey N., Marya S. Rozanova Smith, Andrey K. Krivorotov, Elena M. Klyuchnikova, Valeriy L. Mikheev, Alexander N. Pelyasov, and Nadezhda Yu. Samtatina. "The Russian Arctic by 2050: Developing Integrated Scenarios." *Arctic* 74, no. 3 (2021): 306–22.

Piliasov, Aleksandr, *Arkticheskaia promyshlennost' i promyshlennaia politika*. Moscow, Smolensk: Universum, 2023.

Plotnitskiy, Andrey, and Arnab Roy Chowdhury, "Killing Nature—Killing Us: 'Cultural Threats' as a Fundamental Framework for Analyzing

Selected Bibliography

Indigenous Movements against Mining in Siberia and the Russian North." *Post-Soviet Affairs* 39, no. 3 (2023): 195–212.

Radvanyi, Jean. "When Putin Turned Geographer." *Hérodote*, no. 166 (2017): 113–32.

Rozanova, Marya. "Indigenous Urbanization in Russia's Arctic: The Case of Nenets Autonomous Region." *Sibirica: Interdisciplinary Journal of Siberian Studies* 18, no. 3 (2019): 54–91.

Rozanova-Smith, Marya, Stanislav Ksenofontov, and Andrey N. Petrov, "Indigenous Urbanization and Indigenous Urban Experiences in the Russian Arctic: The Cases of Yakutsk and Naryan-Mar." In *Urban Indigeneities*, edited by Dana Brablec and Andrew Canessa, 209–42. Tuczon: University of Arizona Press, 2023.

Sergunin, Alexander, ed. *Handbook of Research on International Collaboration, Economic Development, and Sustainability in the Arctic*. Chicago: Illinois Global Institute, 2019.

Sergunin, Alexander. "Introduction: Quo Vadis? The Arctic between Nationalism and Globalism." *Polar Journal* 11, no. 1 (2021): 1–10.

Sergunin, Alexander. "Thinking about Forthcoming Russian Arctic Council Chairmanship: Challenges and Opportunities." *Polar Science* 29 (2021): 1–9.

Sergunin, Alexander, and Valerii Konyshev, "Russia's Arctic Strategy," in *Russia Strategy, Policy and Administration*, edited by Irvin Studin, 135–44. London: Palgrave Macmillan, 2018.

Shiklomanov, Nikolai I. "All Fall Down? Urban Infrastructure and Permafrost in the Russian Arctic." *Russian Analytical Digest* 261 (2020): 7–10. https://css.ethz.ch/content/dam/ethz/special-interest/gess/cis/center-for-securities-studies/pdfs/RAD261.pdf.

Shiklomanov, Nikolay, Dmitry Streletskiy, Luis Suter, Robert Orttung, and Nadezhda Zamyatina, "Dealing with the Bust in Vorkuta, Russia." *Land Use Policy* 93 (2020): 1–11.

Siddiqi, Asif. "Atomized Urbanism: Secrecy and Security from the Gulag to the Soviet Closed Cities." *Urban History* 49, no. 1 (2022): 190–210.

Slade, Gavin, Laura Piacentini, and Alena Kravtsova. "Ghosts of the Gulag: Negotiating Spectres of the Penal Past in Northern Russia." *British Journal of Criminology* 64, no. 1 (2024): 17–33.

Slezkine, Yuri. *Arctic Mirrors: Russia and the Small Peoples of the North*. Ithaca, NY: Cornell University Press, 1994.

Sokolov, Denis. "Islam in the Gold Heart of Russia: Ingush Islamic Communities in the Kolyma." *Problems of Post-Communism* 67, nos. 4–5 (2019): 1–14.

Sokolov, Denis. "Ugra, the Dagestani North: Anthropology of Mobility between the North Caucasus and Western Siberia." In *New Mobilities and Social Changes in Russia's Arctic Regions*, edited by Marlene Laruelle, 176–93. Abingdon: Routledge, 2016.

Sorokina, Tatiana, and William G. Phalen. "Legal Problems of the Northern Sea Route Exploitation: Brief Analysis of the Legislation of the Russian Federation." In *International Marine Economy: Law and Policy*, edited by Myron H. Nordquist, John Norton Moore, and Ronán Long, 99–120. Leiden: Brill, 2017.

Sorokina, Tatiana Yu. "Pollution and Monitoring in the Arctic." In *Global Arctic: An Introduction to the Multifaceted Dynamics of the Arctic*, edited by Matthias Finger and Gunnar Rekvig, 229–53. Berlin: Springer Nature, 2022.

Spielman, Brian. "An Evaluation of Russia's Impending Claim for Continental Shelf Expansion: Why Rule 5 Will Shelve Russia's Submission." *Emory International Law Review* 23 (2009): 309–49.

Stepanov, Igor V., and Peter Ørebech. *Legal Implications for the Russian Northern Sea Route and Westward in the Barents Sea*. Oslo: Fridtjof Nansens Institutt, 2005.

Streletskiy, Dmitri A., Luis J. Suter, Nikolay I. Shiklomanov, Boris N. Porfiriov, and Dmitry O. Eliseev. "Assessment of Climate Change Impacts on Buildings, Structures and Infrastructure in the Russian Regions on Permafrost." *Environmental Research Letters* 14, no. 2 (2019). https://iopscience.iop.org/article/10.1088/1748-9326/aaf5e6.

Sukneva, Svetlana, and Marlene Laruelle. "A Booming City in the Far North: Demographic and Migration Dynamics of Yakutsk, Russia." *Sibirica: The Journal of Siberian Studies*, no. 3 (2019): 9–28.

Tchebakova, N. M., E. Parfenova, and A. J. Soja. "The Effects of Climate, Permafrost and Fire on Vegetation Change in Siberia in a Changing World." *Environmental Research Letters* 4, no. 4 (2009): 1–9.

Thompson, Niobe. "Administrative Resettlement and the Pursuit of Economy: The Case of Chukotka." *Polar Geography* 26, no. 4 (2002): 270–88.

Timtchenko, Leonid, "The Russian Arctic Sectoral Concept: Past and Present." *Arctic* 50, no. 1 (1997): 29–35.

Tomaselli, Alexandra, and Anna Koch. "Implementation of Indigenous Rights in Russia: Shortcomings and Recent Developments." *International Indigenous Policy Journal* 5, no. 4 (2014): 1–21.

Vallikivi, Laur. *Words and Silences: Nenets Reindeer Herders and Russian Evangelical Missionaries in the Post-Soviet Arctic*. Bloomington: Indiana University Press, 2024.

Vasiliev, Vladimir, Boris Krasnopolsky, and Alexander Piliasov. *Rozhdennyi ob"edeniat' (k 30 letiu Severnogo Forum*. Moscow-Smolensk: Universum, 2023.

Selected Bibliography

Vasil'kova, N. N., A. V. Evai, E. P. Martynova, and N. I. Novikova. *Korennye malochislennye narody i promyshlennoe razvitie Arktiki.* Moscow: Shchadrinskii dom pechati, 2011.

Vidal, Florian. "Russia in the Arctic: The End of Illusions and the Emergence of Strategic Realignments." Institut Français des Relations Internationales (IFRI), July 2024, https://www.ifri.org/en/papers/russia-arctic-end-illusions-and-emergence-strategic-realignments.

Vylegzhanin, A. N., and V. L. Zilanov. *Spitsbergen: Legal Regime of Adjacent Marine Areas.* Portland, OR: Eleven International, 2007.

Wahden, Lukas B. "Big Words, Small Deeds: Russia and China in the Arctic." Institut de Recherche Stratégique de l'École Militaire, Research Paper no. 141, February 28, 2024, https://www.irsem.fr/media/5-publications/nr-irsem-141-wahden.pdf.

Wall, Colin, and Njord Wegge. "The Russian Arctic Threat: Consequences of the Ukraine War." Center for Strategic & International Studies: CSIS Briefs, January 23, /https://www.jstor.org/stable/pdf/resrep47094.pdf?acceptTC=true&coverpage=false&addFooter=false.

Wallace, Ron. "The Case for RAIPON: Implications for Canada and the Arctic Council." Calgary, Alberta: Canadian Defence and Foreign Affairs Institute, 2013. https://d3n8a8pro7vhmx.cloudfront.net/cdfai/pages/42/attachments/original/1413674425/The_Case_for_RAIPON.pdf?1413674425.

Wilson Rowe, Elana. "Analyzing Frenemies: An Arctic Repertoire of Cooperation and Rivalry." *Political Geography* 76 (2020): 1–10.

Wilson Rowe, Elana. *Russian Climate Politics.* New York: Palgrave Macmillan, 2014.

Wu, Ruonan, Gareth Trubl, Neslihan Taş, and Janet K. Jansson. "Permafrost as a Potential Pathogen Reservoir." *One Earth* 5, no. 4 (2022): 351–60.

Yarlykapov, Ahmet. "Divisions and Unity of the Novy Urengoy Muslim Community." *Problems of Post-Communism* 67, nos. 4–5 (2020): 338–47.

Zamiatina, Nadezhda Iu. "Development Cycles of Cities in the Siberian North." In *The Siberian World*, edited by John P. Ziker, Jenanne Ferguson, and Vladimir Davydov, 352–63. London: Routledge, 2023.

Zamiatina, Nadezhda Iu. "Igarka as a Frontier: Lessons from the Pioneer of the Northern Sea Route." Series: Humanities & Social Sciences. *Journal of Siberian Federal University* 13, no. 5 (2020): 783–99.

Zamiatina, Nadezhda Iu. *Zhiznestoikost' arkticheskikh gorodov: Teoriia, kompleksnyi analiz i primery transformatsii.* Moscow: Izdatel'skie resheniia, 2023.

Zamiatina, Nadezhda, and Ruslan Goncharov. "'Agglomeration of Flows': Case of Migration Ties between the Arctic and the Southern

Regions of Russia." *Regional Science Policy & Practice* 14, no. 1 (2022): 63–85.

Zamiatnika, Nadezhda Iu., and Ruslan V. Goncharov. "Arkticheskaia urbanizatsiia: Fenomen i sravnitel'nyi analiz." *Vestnik Moskovskogo Universiteta: Geografiia*, no. 5 (2020): 69–82.

Zellen, Barry Scott. "The Arctic Council Pause: The Importance of Indigenous Participation and the Ottawa Declaration." Arctic Circle. June 14, 2022. https://www.arcticcircle.org/journal/the-importance-of-indigenousparticipation-and-the-ottawa-declaration.

Zubarevich, Natal'ia. "Four Russias: Human Potential and Social Differentiation of Russian Regions and Cities." In *Russia 2025*, edited by Maria Lipman and Nikolai Petrov, 67–85. New York: Palgrave Macmillan, 2015.

INDEX

Alaska 2, 6, 12, 34, 44, 47, 52, 76, 91, 102
Apatity 58, 81, 88
Arctic
  Circle 3, 15, 19, 34, 76, 80
  Council 1, 24, 34–9, 41, 50, 54, 94, 105
  Ocean 1, 2, 6, 7, 9, 10, 12, 15, 20, 21, 34, 39, 50, 70, 72, 76, 78, 97, 101, 105
  Strategy 39, 44, 53, 67
  Zone 2, 3, 6, 14, 15, 32, 55, 58, 66–70, 73, 79, 81, 82, 85, 86, 103, 105
Arkhangelsk 5, 14, 15, 26, 28, 41, 42, 46, 50, 51, 70–2, 76, 78, 80–6, 90, 101

Barents Sea 7, 9, 10, 28, 40, 42, 47, 64–8, 75, 103, 104
Barentsburg 11, 39, 54
Bering Sea 1, 7–9, 52

Canada 1, 2, 6–8, 12, 21, 34, 45, 47, 67, 68, 70, 71, 91
Cherepovets 59
Chilingarov, Arthur 36, 37
China 29, 39, 46, 47, 50–3, 60, 62, 63, 65, 70, 97, 99
Chukotka 15, 27, 45–7, 68, 69, 72, 75, 80, 82, 83, 85, 86, 92
Cold War 1, 9, 12, 33–5, 41, 44, 45, 55, 60, 61, 77
Crimea 37, 50, 61, 72

Denmark 2, 6, 7, 34, 45, 67
Dolgans 92
Dudinka 28, 59, 79, 81, 96

Eurasia 7, 12
European Union (EU) 23, 38, 62, 64, 66, 70
Evenks 92

Finland 1, 34, 38, 44, 47, 67, 103–5
Franz Joseph Land 24

Gazprom 51, 61, 62, 64, 67, 68, 78, 82, 84, 89, 124
Global South 4, 39, 42, 53, 54, 97
Gorbachev, Mikhail 2, 33, 57
Greenland 2, 5, 12, 33, 34, 40, 45, 47, 70, 71, 91, 98
Grey Zone 9, 10, 103
Gulag 18, 19, 27, 56, 67, 79, 83, 101, 102

Iceland 34, 38, 47, 70
Ivan IV 5

Japan 23, 39, 63, 76, 102

Kamchatka 25, 75, 76, 86
Khanty-Mansi Autonomous District 15, 75, 80, 83, 85, 86, 92, 93, 95
Khodorkovsky, Mikhail 28, 29
Khrushchev, Nikita 60
Kola Peninsula 6, 24, 40, 55, 58, 59, 65, 71, 73, 79

Index

Kolyma 18, 56
Komi 15, 41, 51, 76, 84–6, 90, 92, 94, 95, 103
Krasnoyarsk 15, 25, 27, 37, 81, 86
Kyoto Protocol 23, 104

Labrador 12
LNG (Liquefied Natural Gas) 30, 31, 50, 51, 62–6, 68, 72, 82
Lukoil 28, 56, 64, 67

Magadan 15, 18, 26, 34, 59, 79, 83, 85, 86, 95
Medvedev, Dmitry 10, 14, 104
Mendeleev 7, 8, 57
Modi, Narendra 53
Moscow 3, 6–9, 11, 21–3, 30–3, 35–41, 45–7, 50, 53, 54, 59, 63–5, 67, 72, 73, 77, 83, 91, 93, 97, 99
Murmansk 2, 12, 15, 24, 26–8, 31, 33, 34, 41, 51, 53, 54, 57, 58, 64, 65, 70, 75, 76, 78, 80–6, 88, 90, 101

Nenets Autonomous District 25, 41, 51, 64, 67, 68, 75, 82, 84, 86, 87, 105
Nikolayev, Mikhail 34
Nord Stream 60, 61, 64
Norilsk 18–20, 22, 24–6, 28, 29, 56, 58, 59, 67, 73, 75, 79–82, 85, 88, 89, 101
Nornickel 24, 51, 56, 59, 67, 79, 82, 96
Northern Fleet 1, 40–2, 79, 87
North Atlantic Treaty Organization (NATO) 1, 9, 33, 36, 39, 41, 43–5, 54, 105, 106
Northeast Passage 1, 2, 5, 6, 11, 12, 14, 26
North Korea 54
North Pole 6–7, 12–13, 33, 36, 72, 101–4
Northern Sea Route 2, 4–6, 13, 14, 20, 24, 26–31, 34, 40, 43–7, 50–7, 59, 62, 66, 71–3, 76, 77, 80, 101, 105, 106
Norway 1, 6–11, 34, 38–40, 44, 67, 69, 70, 91, 101, 103
Novatek 29–31, 62–6, 72
Novaya Zemlya 1, 28, 43, 45, 57, 71, 102, 103

Ob 27–9, 56, 60–2, 65, 75, 87, 101
Oslo 9, 11, 39

Pechora 18, 26, 27, 59, 101–3
People of the North 92
Polar Silk Road 47, 73
Putin, Vladimir 1, 15, 22, 23, 27, 29, 36, 37, 40, 45, 46, 53, 56, 57, 62, 64–7, 92

RAIPON (Russian Association of Indigenous Peoples of the North) 93, 94
Rosatom 30, 31, 42, 45, 54, 105
Rosneft 46, 50, 53, 56, 62, 64, 67
Russia 1–14, 21, 23, 25, 29, 31, 33–41, 43–7, 50–4, 60, 62, 64–7, 69, 70, 72, 75, 78, 80, 82, 84, 86, 88–97, 99, 103–5

Sabetta 62–5
Saint Petersburg 23, 44, 46, 52, 73
Sakha (see also Yakutia) 15, 27, 29, 45, 47, 52, 72, 75, 79, 84–6, 92–4, 96
Salekhard 27, 81–4, 95
Scandinavia 1, 21, 71
Sechin, Igor 46, 64
Severnaya Zemlya 28
Severomorsk 1, 41
Shalamov, Varlam 19
Shoigu, Sergei 36, 37, 43

Index

Siberia 2, 4, 6, 7, 14, 20, 24–9, 37, 50, 56, 62, 65, 67, 76, 78, 80, 87, 91, 92, 94, 99, 101, 102
Silk Road 63
Solovki Islands 18
Solzhenitsyn, Alexander 19
South Korea 39, 66, 70
Soviet Union (see also USSR) 1, 2, 6, 8, 10, 12, 14, 20, 23, 33, 42, 57, 59, 60, 78, 87, 88, 91, 101–4
Svalbard 9–11, 24, 38, 47, 54, 101
Sweden 1, 34, 44, 47, 67

Taimyr Peninsula 75, 80, 83
Timchenko, Gennady 62
Total (Gas Company) 63–5
Trump, Donald 2, 9, 45, 98

Ukraine 1, 14, 33, 37, 39, 42–4, 46, 50, 57, 61, 64, 72, 73
United Nations (UN) 7, 39, 88, 93, 103, 104
  Division for Ocean Affairs and the Law of the Sea (UNCLOS) 7, 9, 40, 103–5

United States 1, 2, 6–8, 12, 13, 23, 29, 33–6, 40, 44, 45, 51, 63, 65, 66, 70, 77, 88, 94, 102–14
Ural 25, 60, 68, 84, 101

Vladivostok 13, 28, 29, 31, 34, 44, 53, 65
Vologda 73, 102
Vorkuta 27, 56, 79, 81, 83, 84, 88, 130, 140

White Sea 18, 47, 75, 101
Wrangel Island 7, 24, 43, 72, 103

Yakutia (see also Sakha) 15, 26, 27, 34, 45, 46, 52, 56, 72–5, 86, 92, 94
Yakutsk 22, 27, 78–82, 84, 96
Yamal 2, 29, 37, 47, 50, 62, 65, 75
Yamalo-Nenets District 15, 25, 41, 67, 83, 85–7, 92, 102, 105
Yeltsin, Boris 29, 56, 57, 67
Yenisei 21, 27, 28, 59, 80, 96
Yukos 28, 29, 56